English COMPOSITION WITH INTEGRATED REVIEW

Guided Notebook

HAWKES LEARNING

Editors:
Katherine Cleveland,
Caitlin Edahl,
Sarah Quinn

Creative Director:
Tee Jay Zajac

Designers:
Trudy Gove,
Patrick Thompson,
Tee Jay Zajac

Cover Design:
Natalie Ezabele

VP Research & Development: Marcel Prevuznak

Director of Content: Kara Roché

HAWKES LEARNING

A division of Quant Systems, Inc.

546 Long Point Road
Mount Pleasant, SC 29464

Printed in the United States of America

10 9 8 7 6 5 4 3 2 1

ISBN: 978-1-64277-052-0

Table of Contents

Chapter 4R
Integrated Review: Parts of the Essay

Chapter 5R
Integrated Review: Reading Critically

Chapter 6R
Integrated Review: Writing Critically

Chapter 7R
Integrated Review: Research

Chapter 8R

Integrated Review: Unique Forms of Writing

Chapter 9R

Integrated Review: Basics of Grammar & Mechanics

Chapter 10R

Integrated Review: Grammatical Sentences

Chapter 0R
Integrated Review: Study Skills

Lesson 0R.1
Understanding Different Learning Styles

OBJECTIVE

★ Recognize the characteristics of the eight learning styles.

BIG IDEA

Everyone understands and processes information in different ways. Researchers Richard Felder and Linda K. Silverman have organized learning styles into four pairs:

1.

3.

2.

4.

In this lesson, you will learn about the following learning styles:

- Visual and Verbal Learning
- Active and Reflective Learning
- Sensing and Intuitive Learning
- Sequential and Global Learning

Visual and Verbal Learning

Visual learners use ________________________ to learn new concepts. ______________________, ______________________, and ____________________ are much more memorable to visual learners than words.

List four resources that visual learners might find most useful:

1.

3.

2.

4.

Verbal learners understand information from ____________________. These learners are usually good at remembering what people said, but they may ___________________ to remember what people ______________________ __________________________.

List five resources that verbal learners might find most useful:

1.
2.
3.
4.
5.

Active and Reflective Learning

Active learners like to be _____________________.

Active learners enjoy lively __________________ _______________________ or ________________ ___________________.

Reflective learners like to ___________________ on new _________________________ by themselves.

Define the term *reflecting*.

Which type of learner:

- enjoys working with friends on projects?
- might consider himself a hands-on learner?
- sometimes finds herself procrastinating on big projects?
- sometimes practices important conversations in his head?
- is good at figuring things out on her own?

Sensing and Intuitive Learning

Sensing learners like to solve problems through methods that are ____________________________ and ________________________. They are the most comfortable with objective facts and don't like surprises.

Intuitive learners enjoy ___________________________ all the possibilities in a situation. They often find new and _____________________ ways to solve a ______________________ and feel comfortable working with ________________________ ideas and topics.

Sequential and Global Learning

Sequential learners prefer to learn information in a ______________________ ______________________.

Explain the meaning of *linear*.

Global learners like to understand the _________-___________________ ___________________ ideas first.

Which type of learner:

- might say "I ll come back to that later?"
- prefers to learn something in sequential order?
- may feel overwhelmed by a project unless it's been broken down into orderly tasks?
- might store up knowledge until suddenly everything fits together and makes sense?
- might want someone to describe the big picture for him or her?

Lesson Wrap-up

Key Terms

Define the following Key Terms from this lesson.

Active Learning:

Global Learning:

Intuitive Learning:

Learning Style:

Reflective Learning:

Sensing Learning:

Sequential Learning:

Verbal Learning:

Visual Learning:

Lesson 0R.2

Determining Your Personal Learning Styles

OBJECTIVES

- ★ Apply an understanding of learning styles to real-life situations.
- ★ Create success strategies for different types of learning.

BIG IDEA

Knowing your own personal learning styles can help you overcome any frustration you feel when learning new information.

List the eight different types of learning styles:

1.
2.
3.
4.
5.
6.
7.
8.

In this lesson, you will learn about the following:

- Exploring Your Personal Learning Styles
- Learning Style Strategies
- Learning Styles Outside of School

Exploring Your Personal Learning Styles

Read through the student profiles in the text carefully. Then, for each pair, circle the style you lean toward.

- Visual Learner or Verbal Learner
- Active Learner or Reflective Learner
- Sensing Learner or Intuitive Learner
- Sequential Learner or Global Learner

Learning Style Strategies

Visual and Verbal Learning

If you're a <u>visual</u> learner, you should use more ______________________, ____________________, or ____________________ during your study time.

If you're a <u>verbal</u> learner, look for opportunities to study with ______________ or ______________ recordings.

Active and Reflective Learning

If you're an active learner, you like to turn ____________________ into ________________.

If you're a reflective learner, you usually spend time _______________ about an idea before ___________________ on it.

Sensing and Intuitive Learning

If you're a sensing learner, you feel the most comfortable learning _______________ facts and applying those facts to the ____________________ world.

If you're an intuitive learner, you enjoy thinking _____________________________.

Sequential Learning and Global Learning

If you're a sequential learner, you learn information in_______________, _______________ steps.

If you're a global learner, you learn information in _________________________ bursts of understanding and like to look at the ____________ ________________________ first.

Learning Styles Outside of School

You can use the same learning styles strategies to help you find success at _____________________ or in your _______________________ life.

Lesson Wrap-up

Test Yourself

Next to each statement, write **T** for True or **F** for False. Check your work using the Answer Key in the back of the book.

1. _____ Knowing your individual learning style(s) can help you find success at work or in your personal life.
2. _____ You should not let your instructor know that you might need the information explained in a different way.
3. _____ Learning styles should never be an excuse for bad study habits.
4. _____ Most instructors would be happy to meet with you if you explain you're having some difficulty with the class and you'd like to become a better student.
5. _____ An instructor would not want you to use any other resources besides the textbook and notes for his/her class.

6. ______ Many people are either visual learners or verbal learners, and it is best if they limit themselves to just visuals or plain text.

7. ______ Using a balance of your learning styles will help you make the most of your study time.

8. ______ Global learners can use their knowledge of the big picture to understand smaller details.

9. ______ Learning styles are helpful only for academic purposes.

Choose the best term and fill in the blank. Check your work using the Answer Key in the back of the book.

10. Which kind of learner might act a little too quickly and then not understand the information?
(*reflective* or *active*) ______________________________

11. Which kind of learner enjoys thinking creatively? (*sensing* or *intuitive*) ______________________

12. Which kind of learner might say, "I need a smaller, more manageable task"? (*sequential* or *global*)
____

Key Terms

Define the following Key Terms from this lesson.

Active Learning:

Global Learning:

Intuitive Learning:

Learning Style:

Planner:

Reflective Learning:

Sensing Learning:

Sequential Learning:

Verbal Learning:

Visual Learning:

Workspace:

Lesson 0R.3

Stress Management

OBJECTIVES

★ Learn basic stress-management strategies.
★ Understand the causes of stress.

BIG IDEA

Stress can't always be avoided completely, but strategies for keeping it under control can reduce your likelihood of becoming overwhelmed.

In this lesson, you will learn three ways to manage your stress more effectively:

- Determine Causes of Stress
- Put Everything into Perspective
- Make a Plan to Move Forward

Determine Causes of Stress

Define the term **external stress**.

Define the term **internal conflict**.

Read the three scenarios in the lesson about Jordan, Kimani, and Layla, who are all experiencing stress. Make a list of the **internal** and **external** factors causing stress in these situations. Write your list in the table below.

Student	Internal Factors	External Factors
Jordan		
Kimani		
Layla		

On Your Own

Think through the causes of your own stress and write them in the table below. Make sure to consider both external and internal sources.

External Causes	Internal Causes

Put Everything into Perspective

Take the time to think through each situation ____________________________ and ask yourself some of the following questions:

1.
2.
3.

On Your Own

Prioritize your own stress in the table below.

Stress	Rank	Reason

Make a Plan to Move Forward

Write a brief definition for each stress-management strategy.

1. Take action:

2. Keep yourself organized:

3. Don't be afraid to ask for help:

4. Relax and recharge:

5. Prevent stress before it starts:

Lesson Wrap-up

Test Yourself

Next to each statement, write **T** for True or **F** for False. Check your work using the Answer Key in the back of the book.

1. ______ You should never feel stress at school; if you do, you're probably doing something wrong.
2. ______ The first step in managing your stress is to determine what's causing it.
3. ______ Procrastination means "to get your priorities right."
4. ______ It's best to procrastinate when planning when to study.
5. ______ Planning and organizing are the opposite of procrastinating.
6. ______ Taking action to manage your stress might mean that you will make a to-do list and place the most important tasks at the top.
7. ______ The Student Services Office at your school is a good place to seek out help if you need counseling or tutoring services.

Choose the best answer and fill in the blank. Check your work using the Answer Key in the back of the book.

8. A student who has three tests in one day is experiencing which kind of stress? (*internal* or *external*)

    ____

9. A student who's worried about a science exam because she doesn't want to disappoint her parents would be experiencing which kind of stress? (*internal* or *external*) ______________________

10. Which of the following would be your top priority? (*your semester final exam next week* or *a class project due in a few months*) ____

Key Terms

Define the following Key Terms from this lesson.

External Stress:

Internal Conflict:

Student Services:

Lesson 0R.4

Keeping Yourself Organized

OBJECTIVE

- ★ Learn basic organization strategies.

BIG IDEA

Staying on top of your responsibilities is much easier when you have a structured plan for keeping everything organized.

This lesson will discuss three helpful ways to keep yourself organized and on track:

- Keep a Planner
- Use a File System
- Create a Workspace

Keep a Planner

A **planner** is a place for you to organize your ____________________ and record any ______________ ________________ or ______________________.

You should also gather the following items:

- ______________________________________
- ______________________________________
- ______________________________________
- ______________________________________
- ______________________________________

Explain the top-down strategy:

Checklist: Filling in a Student Planner

What three steps should you take when recording events in your planner?

✓

✓

✓

Define the term **digital planner**.

Use a File System

Why is it a good idea to save class papers?

Create a Workspace

An organized **workspace** is essential for working ______________________________.

Describe your ideal workspace for studying:

Why do you think it is important to keep your workspace free of clutter?

What are some of the materials you should have at your workspace?

On Your Own

To think through your own study habits, answer the following questions:

- Do you prefer studying in the morning or evening?
- When you study, do you prefer silence or noise?
- Would you rather study inside or outside?
- Do you prefer studying alone or in a group?

Lesson Wrap-up

Test Yourself

Next to each statement, write **T** for True or **F** for False. Check your work using the Answer Key in the back of the book.

1. ______ Once you've written in your planner, you should not make any changes.
2. ______ Using a top-down strategy means writing down all non-essential items first.
3. ______ If you have a very good memory, it's probably not necessary for you to use a planner.
4. ______ Two items you should gather for writing down important dates are your class syllabi (plural of *syllabus*) and your work schedule.

5. ______ You can use quizzes and worksheets from your class to study for tests.

6. ______ You should be familiar with the process of saving electronic copies of assignments to folders and sub-folders on your computer.

7. ______ It is okay to have music or noise in your workspace if it helps you concentrate.

8. ______ It's best not to have access to social media sites like Facebook or Twitter while working in your workspace.

9. ______ If clutter doesn't distract you while you're working, then it's not necessary to keep your workspace clutter-free.

10. ______ If you use a room in the library as your workspace, it's important to plan ahead and bring all your supplies and books with you.

Key Terms

Define the following Key Terms from this lesson.

Active Learning:

Digital Planner:

Global Learning:

Learning Style:

Planner:

Reflective Learning:

Syllabus:

Workspace:

Lesson 0R.5

Time Management

OBJECTIVES

- ★ Create a time budget to manage time more effectively.
- ★ Learn basic time-management strategies.

BIG IDEA

Creating a time budget can help you manage your time effectively, thereby reducing your stress levels and boosting your productivity.

This lesson will discuss three strategies for managing your time more effectively:

- Use a Time Budget
- Take Breaks
- Avoid Multitasking

Use a Time Budget

Just like a ____________________ budget shows you how you spend your money, a **time budget** shows you how you spend your ______________________.

Use the following exercises to build your own time budget.

To begin budgeting your time, you first need to get an idea of how you usually spend it. You can do this by creating a record of activities that breaks down your time into increments of 15 minutes.

Here's an example of what your record of activities might look like:

Record of Activities	
Time Block	**Activity**
12:00am–7:30am	Sleeping
7:30am–8:15am	Getting ready for class
8:15am–8:45am	Driving
8:45am–9:00am	Walking to class
9:00am–10:30am	Speech class
10:30am–10:45am	Getting to next class
10:45am–11:30am	English class

You can use the following template to create your own record of activities.

If you can't track your activities over a full-week span, try to record at least 3–5 days' worth. The goal is to ensure that your record represents how you spend your time in a typical week.

Record of Activities	
Time Block	**Activity**

Review the activities you entered and assign each of them to one of the following categories:

- **Sleep**: Time spent sleeping
- **Meal**: Time spent preparing food and eating meals
- **Work**: Time spent at work and any travel time to work
- **Class**: Time spent in class and any travel time to class
- **Study**: Time spent studying and working on class assignments
- **Extracurricular**: Time spent on activities like club meetings, volunteer work, and sports practice
- **Exercise**: Time spent exercising
- **Personal**: Time spent on personal care, laundry, appointments, etc.
- **Family**: Time spent on family commitments
- **Entertainment**: Time spent on hobbies, TV, movies, and the internet
- **Social**: Time spent at parties or with friends
- **Other/Add your own**: Time that doesn't fit into another category

Now, use the table below to add up how many hours you spent on each category. These calculations will help you determine what changes you need to make to budget your time more effectively.

Time Budget Calculator		
Hours	**Minutes**	**Category**

Hours	Minutes	Category
Total:	**Total:**	

Look back at your days of activities and answer the following questions:

1. Do you feel like you had enough time to fulfill all your responsibilities?

2. Were you surprised by how much time was spent on any particular activity?

3. What are some important activities you should have spent more time doing?

4. What are some activities you'd like to spend more time doing in the future?

5. What are some activities you'd like to spend less time doing in the future?

Based on your answers to the previous questions, create a new weekly time budget. Use the following template to prepare your weekly time budget before filling out the next week in your planner.

Week of: ______________________

	SUN	MON	TUES	WED	THURS	FRI	SAT
7AM							
8AM							
9AM							
10AM							
11AM							
12PM							
1PM							
2PM							
3PM							
4PM							
5PM							
6PM							
7PM							
8PM							
9PM							
10PM							
11PM							

Take Breaks

What can be the result of studying as long as possible without a break?

Avoid Multitasking

Define the term **multitasking**.

Is multitasking a time-saver? Why or why not?

Lesson Wrap-up

Test Yourself

Next to each statement, write **T** for True or **F** for False. Check your work using the Answer Key in the back of the book.

_____ 1. A time budget is similar to a financial budget.

_____ 2. When planning your time budget, wait until the end of the week and write down everything you did that week.

_____ 3. After you've recorded all the events in one week, you should categorize each activity.

_____ 4. Not taking breaks and overworking your brain is very important if you want to make it through the semester.

_____ 5. Some students perform better if they can look at Facebook and watch TV while they're studying for their next exam.

Write the correct answers in the spaces below. (Remember, as a general rule, you should set aside at least two hours of study time for every one hour of class time.) Check your work using the Answer Key in the back of the book.

6. In one week (Monday – Friday), Fred spent 4 hours studying, 10 hours with friends at the student center, and 5 hours watching TV. Which activity did he spend too much time doing?

7. Write down the amount of time Fred should devote to studying (Monday–Sunday) if he has a 3-credit-hour class he's supposed to be studying for.

8. If Miguel is taking 12 credit hours this semester, how many hours per week should he devote to studying?

9. What is the name of the strategy which involves writing down the most important activities first in your time budget?

10. Working on several tasks at the same time is called:

Key Terms

Define the following Key Terms from this lesson.

Multitasking:

Planner:

Study Breaks:

Time Budget:

Lesson 0R.6

Study Strategies

OBJECTIVES

- ★ Learn basic study strategies.
- ★ Understand the dangers of procrastination.

BIG IDEA

Practicing good study habits now will help you develop the skills you need to excel in later courses.

This lesson will teach you three ways to improve your study habits:

- Make Studying Part of Your Routine
- Use Creative Study Strategies
- Avoid Procrastination

Make Studying Part of Your Routine

To make studying part of your daily routine, try using the following strategies:

- Find ____.
- Schedule ____.
- Be flexible ____.
- Keep ____.

Use Creative Study Strategies

Everyone has different **learning styles** and ____________________ preferences. Using a variety of creative ____________________ ____________________ will help you find the ones that work best for you.

Checklist: Utilizing Creative Study Strategies

- ✓ Meet ____.
- ✓ Read ____.
- ✓ Use ____.
- ✓ Quiz ____.
- ✓ Use ____.

✓ Create ____________________.

✓ Make up ____________________.

Avoid Procrastination

Define the term **procrastination**.

List some of the factors mentioned in the lesson that cause procrastination and how you might combat them:

- ____________________
- ____________________
- ____________________
- ____________________
- ____________________

Lesson Wrap-up

Test Yourself

Next to each statement, write **T** for True and **F** for False. Check your work using the Answer Key in the back of the book.

1. ______ Studying for a test at the last minute is a good idea because you have less time to forget all the information.
2. ______ You should study at a time of day when you're most likely to be alert.
3. ______ Don't bother studying your notes whenever you have multiple projects due.
4. ______ It's helpful to meet with your instructor and ask how to be successful in the course.
5. ______ Reading your study notes out loud is a waste of time.
6. ______ You should highlight all the text you are studying with just one color.
7. ______ An acronym is an abbreviation using the first letter of each important term.
8. ______ If you procrastinate, you may find that you have more work than you can handle.
9. ______ Fear or lack of self-confidence is not a major factor in causing procrastination.
10. ______ Good time management can help you avoid the trap of procrastination.

Key Terms

Define the following Key Terms from this lesson.

Acronym:

Learning Style:

Planner:

Procrastination:

Lesson 0R.7

Test-Taking Strategies

OBJECTIVE

★ Explore different strategies for reducing test anxiety.

BIG IDEA

Test anxiety can cause you to become overwhelmed by fear or stress, especially during an assessment. Learning to overcome this anxiety will improve your self-confidence as well as your test performance.

This lesson will discuss three strategies for reducing test anxiety:

- Keep a Positive Attitude
- Know What to Expect
- Use Test-Taking Strategies

Keep a Positive Attitude

How does thinking negatively about yourself and your abilities affect your test-taking skills?

To start thinking more ________________________, try these strategies:

- Be more ____.
- Praise ____.
- Think____.
- Find ____.

Know What to Expect

Pre-Test Checklist

✓ Meet ____.

✓ Take ____.

✓ Review ____.

✓ Study ____.

Use Test-Taking Strategies

Test-Taking Checklist

✓ ______________________________

✓ ______________________________

✓ ______________________________

✓ ______________________________

✓ ______________________________

Lesson Wrap-up

Test Yourself

Next to each statement, write **T** for True or **F** for False. Check your work using the Answer Key in the back of the book.

1. ______ Tests are a complete picture of every single one of your abilities.
2. ______ You might encounter tests when applying for certain jobs.
3. ______ Speaking and thinking negatively about yourself could affect your test-taking skills because you will lack confidence.
4. ______ Self-confidence plays a large role in how well you perform on tests.
5. ______ Focusing on when you have achieved your goals does not help when preparing for tests.
6. ______ You should think of setbacks as opportunities for growth.
7. ______ Scrolling through Instagram is a great test-prepping ritual.
8. ______ If possible, you should meet with your instructor if you're nervous about a test.
9. ______ A good strategy for preparing for a test would be to look at old quizzes and tests.
10. ______ Your instructor would never let you study in the testing room or classroom.

Key Terms

Define the following Key Terms from this lesson.

Acronym:

Test Anxiety:

Lesson 0R.8

Taking Advantage of Campus Resources

OBJECTIVE

★ Identify useful campus resources and services.

BIG IDEA

To get the most value out of your education and boost your academic success, take advantage of all the campus resources offered to you.

In this lesson, you will learn the purpose and services of five common campus organizations:

- Student Services
- The Library
- The Academic Success Center
- The Writing Center
- Disabilities Services

Student Services

Student services is the office responsible for ____________________ ________________________.

What are some questions you might want to discuss with your advisor?

1.

2.

3.

4.

5.

The Library

Name a few of the services the library has to offer:

1.

2.

3.

4.

5.

Define the term **LibGuide.**

The Academic Success Center

The **Academic Success Center** exists to help you ______________ ______________________.

Name two ways in which the Academic Success Center can help you:

1.

2.

The Writing Center

The **Writing Center** offers ____________________ ____________________ for all stages of the ________________ ________________.

To get the most out of your time in the Writing Center, follow these suggestions:

Checklist: Writing Center Visit

✓ ____

✓ ____

✓ ____

✓ ____

Disabilities Services

Disabilities Services assists students who have ______________ ______________ or ______________ ______________.

These **accommodations** might include:

- ____
- ____
- ____

Lesson Wrap-up

Test Yourself

Name the office or resource on campus you would use for each scenario. Check your work using the Answer Key in the back of the book.

1. You need to meet with your advisor:
2. Help in finding particular books or resources:
3. If you need accommodations such as extra time on tests or projects:
4. If you want to discuss career resources:
5. Help in writing a thesis statement for your research paper:
6. If you want to schedule an appointment with a tutor:
7. If you want to laminate a class project:
8. You need help with a short story for a creative writing class:
9. Borrowing a video camera:
10. You need to take tests in a quiet environment:

Key Terms

Define the following Key Terms from this lesson.

Academic Success Center:

Accommodations:

Disabilities Services:

LibGuide:

Student Services:

Writing Center:

Lesson 0R.9

Writing with Technology

OBJECTIVES:

- ★ Identify and implement file organization strategies.
- ★ Identify and use shortcut keys.
- ★ Identify the best strategies for saving electronic documents.
- ★ Learn strategies for writing with technology.
- ★ Recognize proper use of spell-check.
- ★ Recognize the steps of page formatting.

BIG IDEA

Learning to utilize technology in an efficient way will help you avoid unexpected stress and save valuable time and energy when writing.

This lesson will discuss four important aspects of writing with technology:

- Saving Your Work
- Using Grammar and Spell-Check
- Formatting the Page
- Using Keyboard Shortcuts

Saving Your Work

When saving your work:

- ▶ Don't wait ______________________________.
- ▶ Make a habit of saving your work every ______________________________.
- ▶ Remind yourself with a ______________________________.
- ▶ Consider saving your work in ______________________________.
- ▶ Think about purchasing an ______________________________.

Using Grammar and Spell-Check

Spell-check should never replace ______________ ______________.

To enable grammar and spell-check in Microsoft Word:

1. Open ______________________________
2. Select ______________________________

3. In the pop-up box, select ____

4. Check ____

Formatting the Page

List the three common formatting options and a brief summary of each:

1.

2.

3.

Using Keyboard Shortcuts

Complete the following table with keyboard shortcuts. Keep the list handy until you get the hang of them.

Command	Windows	Mac
Save your work		
Undo your last action		
Highlight all text		
Cut highlighted text		
Copy highlighted text		
Paste cut/copied text		
Bold highlighted text		

Lesson Wrap-up

Key Terms

Define the following Key Terms from this lesson.

Editing:

Format:

Header:

Page Margin:

Page Number:

Chapter 1R
Integrated Review: Why We Write

Lesson 1R.1
Understanding Purpose

OBJECTIVES

- ★ Recognize the expectations and conventions of different genres.
- ★ Understand the purposes of a longer text.

BIG IDEA

It's important to be able to write for a variety of different purposes, genres, and environments.

Define the term **genre**.

In this lesson, you will learn three steps:

- Determine Your Purpose for Writing
- Choose an Appropriate Genre
- Think about Expectations and Conventions

Determine Your Purpose for Writing

Define the following common purposes for writing:

To inform:

To persuade:

To reflect:

To entertain:

Choose an Appropriate Genre

Genres of writing can be divided into three main groups:

1.

2.

3.

Complete the following table with genres appropriate for each writing environment.

Academic Writing	Professional Writing	Everyday Writing

Think about Expectations and Conventions

After you have selected your genre, you must also consider the __________________________ and __________________________ that are part of that genre.

Tone

Define the term **tone**.

What kind of tone would you expect to encounter in the following texts? Add your own example in the last row.

Example	Tone
Newspaper article	
Blog post about a favorite hobby	
A letter to a manager of a restaurant after experiencing bad service	

Style

In writing assignments, you will generally use a formal style:

- You are expected to sound ______________________ or ____________________.
- You should use more _______________ sentence structure and ______________ terms.
- You should make your writing _________________ and ________________ to understand.

Grammar

In academic writing, you are expected to follow __________________ grammar and __________________ rules.

When you write in everyday genres like emails or text messages, your grammar can be more __________________.

Lesson Wrap-up

Test Yourself

Next to each text, write **I** for Informal or **F** for Formal. Check your work using the Answer Key in the back of the book.

1. _____ Blog post
2. _____ Newspaper article
3. _____ Academic paper
4. _____ Journal entry
5. _____ Published article

Key Terms

Define the following Key Terms from this lesson.

Academic Writing:

Audience:

Complexity:

Contraction:

Everyday Writing:

Formality:

Genre:

Prior Knowledge:

Professional Writing:

Purpose:

Simple Sentence:

Slang:

Tone:

Chapter 2R
Integrated Review: Modes of Writing

Lesson 2R.1

Locating Key Information

OBJECTIVES

- ★ Locate key information in sentences, paragraphs, and essays.
- ★ Understand and identify the topic and main idea of a text.

BIG IDEA

In order to understand the writing mode of a text, there is some key information you should be able to locate in any text—no matter the form it takes.

For example, all texts have ____________________, or subjects. Additionally, you should generally be able to find a big idea that shapes the direction of the text.

In this lesson, you will learn strategies for finding key information in both short and long pieces of writing:

- Finding Key Information in a Sentence
- Finding Key Information in a Text

Finding Key Information in a Sentence

Subject

Define the term **subject**.

Write two sentences and circle the subject in each:

1.

2.

Predicate

Define the term **predicate**.

Write two sentences and circle the predicate in each:

1.

2.

Other Sentence Components

What additional parts of speech can be used to support and/or describe the subject and predicate in a sentence?

Finding Key Information in a Text

Define each of the following components, or parts, of a text:

Topic:

Main Idea:

Supporting Details:

What two questions can you use to identify the topic of a text?

1.

2.

List the questions you should consider to determine the main idea of a text:

1.

2.

3.

In some forms of writing, the main idea is a claim that an author wants readers to understand or believe.

Write your own claim for each of the following topics:

Topic	Claim
Price of school lunches	

Homework in college	
Raising taxes on families	

READING APPLICATION

Read the passage *Building a Safer School Environment* and identify its topic and main idea.

Topic	Main Idea

Topic Sentence

A well-written topic sentence must meet several requirements:

- ____________________
- ____________________
- ____________________

Lesson Wrap-up

Key Terms

Write the number of each term next to its definition.

1. Topic	_____ A word or word group that can be linked to a broad range of specific ideas and details
2. Main Idea	
3. General Information	_____ The general subject of a text
4. Specific Information	_____ A word or word group that introduces a new idea and/or shows the connection between two ideas
5. Signal/Signpost Word	
	_____ The statement or argument that an author tries to communicate
	_____ A word or idea with a narrow focus

Define the following Key Terms from this lesson.

Paragraph:

Predicate:

Purpose:

Subject:

Supporting Detail:

Topic Sentence:

Lesson 2R.2

Classifying Major and Minor Details

OBJECTIVES

★ Recognize supporting details in a text.
★ Distinguish between major and minor details.

BIG IDEA

Supporting details play an important role in the effectiveness of a text. They are used to prove and/or build upon the author's main idea.

In this lesson, you will learn how to locate and utilize supporting details as a reader and as a writer.

- Recognizing Supporting Details
- Sorting Details

Recognizing Supporting Details

Why is it important to use supporting details in writing?

What are **signal words**?

Major Details

What do **major details** support?

What signal words often introduce major details?

Minor Details

What do **minor details** support?

What signal words often introduce minor details?

Sorting Details

Outlining can be a helpful way of visualizing how details relate to each other and to main ideas.

What details would you include in the following pieces of writing? Fill in the outline provided.

1. A paragraph describing your favorite city to visit
    - Favorite city:
        - Major detail:
            - Minor detail:
        - Major detail:
            - Minor detail:
2. A set of directions to your favorite restaurant
    - Destination:
        - Major detail:
            - Minor detail:
        - Major detail:
            - Minor detail:
        - Major detail:
            - Minor detail:
        - Major detail:
            - Minor detail:

Lesson Wrap-up

READING APPLICATION

Read the passage *Pause Before You Post: Social Media Etiquette* and answer the following questions.

1. What is the topic?

2. In the second paragraph, what transition phrase signals the underlined minor detail?

3. Identify one major detail in paragraph 4.

Key Terms

Define the following Key Terms from this lesson.

Main Idea:

Major Details:

Minor Details:

Supporting Detail:

Topic:

Topic Sentence:

Lesson 2R.3

Identifying Organizational Patterns

OBJECTIVE

★ Identify different types of organizational patterns.

BIG IDEA

Knowing the organizational pattern of a paragraph or other written text can help you understand the author's train of thought.

Define the term **organizational pattern**.

This lesson will describe six types of organizational patterns:

- Cause and Effect
- Chronological
- Compare and Contrast
- Order of Importance
- Spatial
- Topical

Cause and Effect

Complete the following table with an effect for each cause. For example, if it is snowing (cause), the roads will be slippery (effect).

Cause	Effect
The baby is hungry.	
It has rained for five days.	

Complete the following table with a cause for each effect.

Cause	Effect
	No one wants to eat at that restaurant.
	People are cheering outside of a church as a couple walks out.

Write three **signal words** or phrases that show *why* or *how*:

1.

2.

3.

READING APPLICATION

Read the passage about revolutionary France and think about how the organization and word choice indicate cause and effect. Then, answer the questions below.

What is the cause given in the first paragraph?

Provide one positive and one negative effect found in the passage.

Positive Effect: **Negative Effect:**

Chronological

When a text is arranged ______________________________, the author discusses the ideas or events in the order that they __________________________.

When you're reading a chronological text, look for signal words that indicate ________________.

List three signal words or phrases that indicate time:

1.
2.
3.

READING APPLICATION

Read the passage about the role of sports in Europe. How does the author indicate the passage of time?

Compare and Contrast

The **compare and contrast** organizational pattern is used to show the ______________________ and ______________________ between two topics.

List three compare and contrast signal words or phrases below:

1.
2.
3.

READING APPLICATION

Read the essay about pet ownership. Then, complete the following chart using examples from the passage.

Owning Cats and Dogs: Similarities	Owning Cats and Dogs: Differences

List three compare and contrast signal words or phrases in the passage:

1.
2.
3.

Order of Importance

When a text is organized by **order of importance**, the information is arranged from ____________________ ______________________________ to ____________________ ______________________________ or vice versa.

List three signal words or phrases that show importance:

1.
2.
3.

READING APPLICATION

Read the essay about classroom size. Then, answer the following questions:

What is the topic of this essay?

What signal phrase is used at the beginning of paragraph 2?

According to paragraph 2, what is an important reason for reducing class size?

Is this essay organized from most important reason to least important or from least important to most important? How do you know?

Spatial

Spatial order is used to describe a subject's ____________________ _________________________________.

A spatial text will use words that show _______________________________.

Write three spatial signal words or phrases:

1.

2.

3.

Use spatial signal words to describe your current location in two or three sentences.

READING APPLICATION

As you read the passage about Ron visiting his childhood home, look for indicators of space, location, and direction. List two examples that you find:

1.

2.

Topical

Topical organization is a general organizational pattern used for _____________________________ _________________________________ main points.

List three signal words used in topically organized texts:

1.

2.

3.

READING APPLICATION

Read the passage about mosquito repellants and identify its topic and main idea.

Topic:

Main Idea:

Lesson Wrap-up

Test Yourself

Match each sentence with the organizational pattern it reflects. Check your work using the Answer Key in the back of the book.

a) Cause and Effect
b) Chronological
c) Compare and Contrast
d) Order of Importance
e) Spatial

_____ 1. There are several major differences between buying a home and renting one.

_____ 2. To the left, I saw food wrappers, and to the right, I saw a broken lamp; clearly, the kids don't keep their space clean when I am out of town!

_____ 3. This recent hurricane had many devastating effects on south Florida.

_____ 4. The American Civil War began in 1861 and ended in 1865.

_____ 5. There are many important issues to consider when selecting a pet for your family.

Key Terms

Define the following Key Terms from this lesson.

Organizational Pattern:

Paragraph:

Signal/Signpost Word:

Topical:

Chapter 3R
Integrated Review: The Writing Process

Lesson 3R.1
Writing a Paragraph

OBJECTIVE

★ Identify the stages in writing a paragraph.

BIG IDEA

Using the academic writing process will ensure that you write a focused paragraph.

The **academic writing process** breaks up an assignment into five stages:

1. 2. 3. 4. 5.

This lesson will review the academic writing process for paragraphs.

Pre-writing

Goals of Pre-writing

What are the questions you answer during the pre-writing stage?

1.

2.

3.

On Your Own

In the following example, is Mario pre-writing? Check the box next to your answer.

> Mario reviews the guidelines given to him by his history instructor for a major paper. He then brainstorms possible topics for his history paper on the Civil War. After that, he is going to the library to search the databases for information about the Civil War.

☐ Yes ☐ No

Drafting

While you are drafting:

- Don't worry too much about ______________________ or ______________________.
- Focus on the ___________________ that you want to cover.
- You can always make ____________________ and _________________________ later.

Revising

The revising stage is usually broken into two parts:

1. Revising for ________________
2. Revising for ________________

Editing

This stage involves fixing ______________, ______________, and ________________ issues.

On Your Own

In the following example, is the advice of Ingrid's friend correct? Check the box next to your answer.

> Ingrid is having trouble finding the errors in a press release for her marketing class. Her friend suggests re-reading her paragraph several times while looking for a different type of writing error each time.

☐ Yes ☐ No

Submitting

During the submitting stage, you will _______________ your paragraph.

- You might format the text in _____________ (Modern Language Association) style. MLA style is a method for ______________________ and _____________________ academic writing.

Lesson Wrap-up

Test Yourself

Next to each statement, write **T** for True or **F** for False. Check your work using the Answer Key in the back of the book.

1. _____ All paragraphs must be eight to ten sentences long.
2. _____ If the ideas in a paragraph aren't focused, the audience won't be able to follow them.

3. ______ There are six stages in the academic writing process.

4. ______ It's okay to skip a step to save time.

5. ______ You can move back and forth between different steps of the writing process.

6. ______ Drafting is the first step of the academic writing process.

7. ______ You can skip the pre-writing stage.

8. ______ Pay attention to grammar and spelling during the drafting stage.

9. ______ Drafting involves writing support sentences and a concluding sentence.

10. ______ Revising is generally broken into two parts: content and organization.

Key Terms

Define the following Key Terms from this lesson.

Academic Writing Process:

Drafting:

Editing:

Main Idea:

MLA:

Organizational Pattern:

Paragraph:

Pre-writing:

Purpose:

Revising:

Submitting:

Support Sentence:

Topic Sentence:

Lesson 3R.2

Writing a First Draft

OBJECTIVE

★ Understand the importance of a first draft.

BIG IDEA

The first step in the drafting stage is to write your first draft. Having a set strategy will help you move more easily through the drafting process.

During the drafting stage, you should focus on ______________________________ between your ideas. There's no pressure to produce an error-free paper. You can always **revise** and **edit** it later.

This lesson will teach you strategies for writing your first draft.

Preparation

To make sure that your writing time is productive, gather all of your __________________________ and ______________________________ ahead of time.

Your **workspace** can also play a big role in your ability to __________________________.

Writing

List the four recommendations you should keep in mind as you write:

1.

2.

3.

4.

Lesson Wrap-up

Test Yourself

Next to each statement, write **T** for True or **F** for False. Check your work using the Answer Key in the back of the book.

1. ______ Spelling and grammar are the main things to focus on when writing your first draft.

2. ______ You must write your introduction first and the conclusion last.

3. ______ You should have a large chunk of writing done by the time you start your first draft.

4. _______ If you get stuck on a sentence or paragraph, push yourself to keep moving forward.

5. _______ You should be clearly supporting your thesis statement in every paragraph.

6. _______ You shouldn't be adding supporting details in the drafting stage.

7. _______ Your workspace plays a big role in your ability to focus.

8. _______ Drafting is the last stage before you turn in your paper.

9. _______ Your first draft doesn't need to be perfect.

10. _______ You should review your outline before you start drafting.

Key Terms

Define the following Key Terms from this lesson.

Audience:

Conclusion:

Drafting:

Editing:

Introduction:

Outline:

Revising:

Supporting Detail:

Thesis Statement:

Lesson 3R.3

Refining a Paragraph

OBJECTIVE

★ Learn a process for revising and editing a paragraph.

BIG IDEA

Revising helps you focus on content, not grammar. Once you have revised the ideas in your paragraph, you can start editing.

In this lesson, you will learn about the third and fourth stages of the writing process:

- Revising Ideas
- Editing Words and Sentences

Revising Ideas

Focus

A **focused** paragraph clearly communicates its __________________ __________________ and uses every sentence to __________________ and __________________ that idea.

On Your Own

Read the following paragraph and identify the sentence that should be deleted to make the paragraph more focused.

> When I was growing up, my brother was the most important person in my life. My brother was important to me for three reasons. First, he was my best friend. We did everything together; we even went to summer camp together. Second, he could already drive. I wouldn't have been able to go anywhere if he wasn't willing to take me. I couldn't wait to get my own license. The third reason my brother was important is that he was the drummer in my band. I could sing, but he was the one who kept us on-beat. My brother played a vital role in my childhood.

List the four questions you can ask yourself as you revise for focus:

1.

2.

3.

4.

Development

A well-developed paragraph presents information in an __________________ way and includes plenty of __________________ to support its ____________ ____________.

On Your Own

Take a look at this paragraph from earlier in the lesson. The highlighted sentence has been added; does it strengthen the paragraph's development? Check the box next to your answer.

> When I was growing up, my brother was the most important person in my life. My brother was important to me for three reasons. First, he was my best friend. We did everything together; we even went to summer camp together. Second, he could already drive. My brother took me to school every morning and band practice every afternoon. I wouldn't have been able to go anywhere if he wasn't willing to take me. I couldn't wait to get my own license. The third reason my brother was important is that he was the drummer in my band. I could sing, but he was the one who kept us on-beat. My brother played a vital role in my childhood.

☐ Yes ☐ No

What questions should you ask yourself as you revise for development?

1.
2.
3.
4.
5.

Revision Strategies

List the strategies for effective revision:

1.
2.
3.
4.

Editing Words and Sentences

Editing involves proofreading your work to make sure the writing is completely free of

________________, __________________, and ________________ errors.

On Your Own

Locate five errors in the following paragraph:

> In mathematics, knowing weather to calculate the area or perimeter of a shape is essential. Calculate the area of a shape. When you want to determine the number of square units the shape occupies. In real life, area was used in situations such as calculating the amount of flooring needed for a new home. Perimeter, on the other hand, is used to calculate the length of the boundary of a shape. For example, perimeter is used too calculate the amount of fencing needed to enclose a playground. Consider the context of the situation before making the choice between area or perimeter.

Lesson Wrap-up

Key Terms

Define the following Key Terms from this lesson.

Academic Writing Process:

Concluding Sentence:

Development:

Drafting:

Editing:

Focus:

Main Idea:

Organizational Pattern:

Purpose:

Revising:

Supporting Detail:

Topic Sentence:

Transition:

Lesson 3R.4

Strengthening Sentences

OBJECTIVES

- ★ Identify the characteristics of complete sentences.
- ★ Understand the characteristics and purposes of dependent and independent clauses.
- ★ Understand the characteristics and purposes of interjections.
- ★ Understand the difference between dependent and independent clauses.
- ★ Recognize subjects and verbs in complete sentences.
- ★ Recognize dependent and independent clauses in sentences.

BIG IDEA

The ability to group words together in complete sentences is an essential part of the writing process.

In this lesson, you will learn how to strengthen your writing by correctly using the following sentence components:

- Capitalization and Punctuation
- Parts of Speech
- Subject and Predicate
- Complete Thought

Capitalization and Punctuation

Complete sentences always start with a ____________ ____________ and end with a ____________, a ____________ ____________, or a(n) ____________ ____________.

Period:

Question mark:

Direct question:

Indirect question:

Exclamation point:

Types of Statements

Declarative sentence:

Interrogative sentence:

Imperative sentence:

Exclamatory sentence:

Parts of Speech

Nouns and Pronouns

A **noun** is a word that represents a ____________, ____________,
____________, ____________, or ____________.

A **pronoun** is a part of speech used to replace a ____________ or another ____________.

On Your Own

Identify the nouns in the following sentences:

> Ava's favorite restaurant has an extensive menu. Ava always orders pho, and her mother, Starla, prefers pad thai.

Identify the pronouns in the following sentences:

> Did you receive the package I sent you? I talked to a FedEx employee, Harriet, and she mentioned that the package may have gotten lost.

Verbs

Verb Types

Action verb:

Linking verb:

Helping verb:

Verb Tenses

Past-tense verb:

Present-tense verb:

Future-tense verb:

Adjectives and Adverbs

Adjectives are used to ____________ ____________ or ____________.

Adverbs ____________ ____________, ____________, or other ____________.

Interjections

Interjections are ____________ or short ____________ of words that show
____________ or ____________.

Subject and Predicate

Write a complete sentence of your own. Circle the subject and underline the predicate.

Complete Thought

When a group of words expresses a **complete thought,** the group of words is ______________________ ______________________. If a group of words contains a subject and a verb but does not express a complete thought, it is known as a ______________________ ______________________.

On Your Own

Read the following paragraph and identify the incomplete sentence.

> While the availability of electronic resources has opened up new opportunities for reducing the cost of education. In first-year composition classes, instructors are selecting electronic textbooks or course companion websites instead of traditional print textbooks. Educators, however, are still unsure about the advantages of either type of text. More research must be conducted to determine if different textbook formats help students learn more effectively.

Lesson Wrap-up

Test Yourself

In the following sentences, add the correct punctuation. Check your work using the Answer Key in the back of the book.

1. Whenever the four of us go shopping, I always spend too much money
2. Did you know the man down the street who passed away
3. Marcus and Steve joined our group and helped us raise money
4. Every time you get behind the wheel of a car, put your seatbelt on
5. I'm *shocked* by how much you look like Steve Carrell

Key Terms

Beside each definition, write the corresponding term.

______________________: A sentence that communicates emotion or surprise

______________________: A word that describes a noun or a pronoun

________________________: A verb tense used to describe events or actions that have not yet taken place or to describe plans or instructions

________________________: A word that takes the place of a noun in a sentence

________________________: A part of speech that describes a verb, adjective, or another adverb

________________________: A punctuation mark that conveys emphasis or strong feeling

________________________: A sentence that makes a general statement that informs the reader

________________________: A sentence that asks the audience a question and usually ends with a question mark

________________________: When a group of words is logically finished

________________________: A word that changes the form of the main verb so that it grammatically fits in the sentence

________________________: A sentence that makes commands and often addresses the implied subject

________________________: A word group that contains a subject and a verb and expresses a complete thought

________________________: A sentence that tells the audience about a question and usually ends with a period

________________________: A word or group of words that adds emphasis or emotion

________________________: A sentence that asks a question

________________________: A verb that links the subject to a description

________________________: A verb tense used to report on an event or reflect on a past experience

________________________: A punctuation mark used to end sentences that make a statement or relay information

________________________: A punctuation mark used at the end of a sentence that asks a question

________________________: The person, place, thing, event, or idea a sentence is about

________________________: A word that represents an action, relationship, or state of being

Lesson 3R.5

Proofreading Strategies

OBJECTIVES

★ Learn a process for proofreading for grammar.
★ Proofread for grammar errors involving punctuation.

BIG IDEA

Being able to correct your own writing errors will help you present your ideas in a professional and academic way.

In this lesson, you will learn about the following strategies:

- Make a List of Your Common Mistakes
- Proofread in Stages
- Try Multiple Reading Techniques
- Take Advantage of Technology
- Take Frequent Breaks

Make a List of Your Common Mistakes

List your most common grammar/spelling mistakes:

Proofread in Stages

Proofreading Checklist: Grammatical Sentences

✓ Sentence Structure:

✓ Agreement:

✓ Punctuation:

✓ Spelling:

Try Multiple Reading Techniques

Name three proofreading strategies that might help you focus on grammar and spelling mistakes:

1. 2. 3.

Take Advantage of Technology

List two ways you can take advantage of technology while you proofread:

1.

2.

On Your Own

Read the sentence below and identify the misspelled word.

Gaming is now headed into the new teritory of virtual reality.

Take Frequent Breaks

______________ to ______________ minutes should be plenty of time for you to stretch your legs or get a drink. Don't allow yourself to take too much time, however, as you might become ____________________.

Lesson Wrap-up

Key Terms

Define the following Key Terms from this lesson.

Comma Splice:

Dependent Clause:

Fused Sentence:

Independent Clause:

Pronoun-Antecedent Agreement:

Run-on Sentence:

Semicolon:

Subject-Verb Agreement:

Lesson 3R.6

Finalizing a Paragraph

OBJECTIVE

★ Understand how to submit a paragraph.

BIG IDEA

Submitting your work is an opportunity to share your ideas with your audience and reflect on all of the hard work that went into your writing.

In this lesson, you will learn two steps for submitting a paragraph:

- Choose a Format
- Print or Upload Your Work

Choose a Format

Format can involve any of the following details:

-
-
-
-
-
-

On Your Own

Read the following writing prompt and identify the sentences that describe formatting requirements.

> A term limit is a law that sets the maximum number of times an individual may hold an elected office. For example, the president of the United States cannot serve more than two four-year terms. People hold varying opinions on term limits; some believe that term limits prevent political corruption while others think they prevent politicians from thinking about long-term issues. Research the history of term limits to better understand the potential positive and negative effects. Then, write an essay explaining your position on the topic. Your paragraph should be double-spaced and in 12-point Times New Roman font. Remember to set page margins at one inch on all sides.

Print or Upload Your Work

Print your work at least ______________ ______________ before the due date.

Look over your printed copy one last time before ______________________________ it.

If you email your writing, double check any ______________________ to make sure you are using the correct file format.

Before sending your document, ______________ it to yourself so that you can ______________ the file and make sure it opens without any issues.

Check your Sent folder to ________________ that the email was delivered without any problems.

On Your Own

Why is it important to print, submit, and/or email an assignment at least one day before it is due? Identify the correct answer.

a) It gives the instructor extra time to read the assignment.

b) Students who turn work in early receive higher grades.

c) It gives you extra time in case you have technology issues.

Lesson Wrap-up

Key Terms

Define the following Key Terms from this lesson.

Academic Writing Process:

Audience:

Editing:

Font:

Format:

Page Margin:

Paragraph:

Purpose:

Revising:

Submitting:

Writing Prompt:

Chapter 4R
Integrated Review:
Parts of an Essay

Lesson 4R.1

Sorting General and Specific Information

OBJECTIVES

- ★ Understand the differences between general and specific information.
- ★ Distinguish between general and specific information in a paragraph.
- ★ Distinguish between general and specific information in an essay.

BIG IDEA

Recognizing general and specific information is a helpful starting point for understanding the parts of an essay and how they work together.

In this lesson, you will learn about the following:

- General and Specific Words & Word Groups
- General and Specific Components of a Text

General and Specific Words & Word Groups

On Your Own

Add to the table with your own examples of general and specific words.

General	Specific
states	Florida, Tennessee, South Carolina
dogs	German shepherd, basset hound, Welsh corgi

Write three phrases or sentences that provide specific details about your dream house:

1.
2.
3.

General and Specific Components of a Text

In a paragraph, sentences work together to ________________ and ________________ a general idea.

List the components of a paragraph in order from general to specific:

READING APPLICATION

You will encounter the following passage in Chapter 4. To prepare, read it now and follow the directions to practice locating general and specific information in a paragraph.

> In the lower forty-eight states, we leave the gate only to sit for hours on the hot tarmac, endure invasive searches with sweaty shoes in hand, and cram our three-ounce shampoo bottles into plastic bags. The boredom, inane rules, and indignities of commercial flight have extinguished the glamor and adventure of air travel. We've lost the sense of wonder that aviation once held. We find no delight in the skies, and even our children are too jaded to marvel that tons of metal can rise up from the earth and soar. But here it is different. To fly over glaciers and desolate mountains is to feel both powerful and small. To fly in this place is to feel alive but afraid. This place is Alaska, where harsh landscapes, violent weather conditions, and vast distances without radio communication make aviation more exciting and deadly than anywhere else in the United States.

After reading this passage once for comprehension, follow these instructions:

1. Read the passage a second time with general and specific information in mind.
2. Circle, underline, or highlight the sentence containing the most general information.
3. Fill in the mind map provided to illustrate the relationships between the topic and details in the passage.

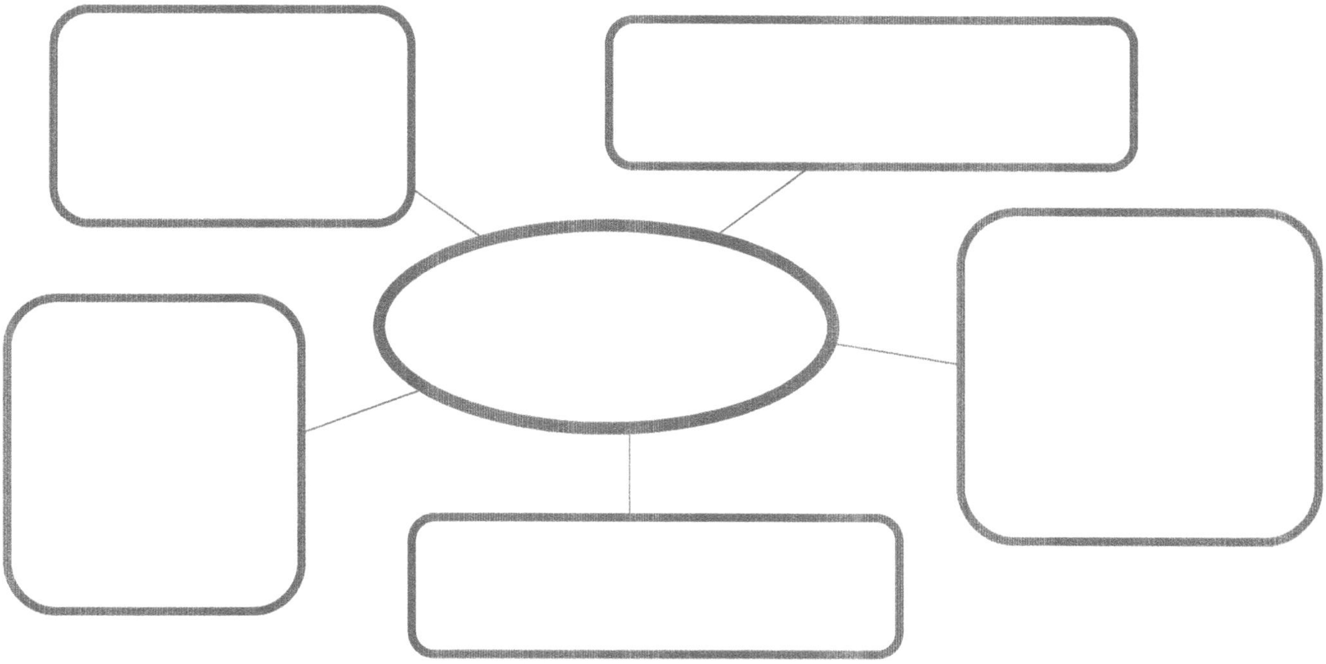

Like ______________________________, essays include both general and specific ideas. The overarching ________________ or goal of an essay influences the ideas in each of its paragraphs.

List the components of an essay in order from general to specific:

Lesson Wrap-up

Key Terms

Define the following Key Terms from this lesson.

Conclusion:

Introduction:

Main Idea:

Mind Map:

Supporting Detail:

Thesis Statement:

Topic:

Topic Sentence:

Lesson 4R.2

Determining Paragraph Focus

OBJECTIVES

★ Identify strategies for choosing a topic.
★ Understand the importance of narrowing the scope of a topic.

BIG IDEA

Carefully considering the topic of a paragraph will ensure that your writing is clear and focused.

Define the term **topic**.

This lesson will teach you three useful steps in choosing a topic and narrowing your scope:

- Review the Assignment Directions
- Use Brainstorming Strategies
- Think about Audience, Purpose, and Constraints

Review the Assignment Directions

Define the term **writing prompt**.

What four questions should you ask yourself when reviewing assignment directions?

1.

2.

3.

4.

On Your Own

Read the following prompts, paying close attention to the assignment guidelines. Then, in the following space, write your own prompt for an assignment in one of your classes.

Prompt 1:

> Write a well-developed paragraph about your learning style. Make sure you have a topic sentence and supporting details. The writing rubric located in the Course Information tab will be used to grade the paragraph. Turn in the typed paragraph on Monday.

Prompt 2:

> Enforcing term limits is a procedure where the number of terms an individual may hold an elected office is set by law. For example, the President of the United States may not serve more than two terms. Research the issue to better understand the potential positive and negative effects of term limits. Then, write an argumentative, five-paragraph essay explaining your position and supporting it with information from scholarly articles. Use MLA style.

On Your Own

In the following example, is Juan's friend correct? Check the box next to your answer.

> Juan is having trouble understanding a writing assignment in his history class. Juan's friend suggests highlighting any special guideline as he re-reads the writing prompt.

☐ Yes ☐ No

Some key words are often used in writing prompts. Define each of these key action words:

Analyze:

Define:

Discuss:

Evaluate:

Identify:

Prove:

Share:

Summarize:

On Your Own

Identify the key action words in the following prompts:

1. Several styles of parenting have been covered in class. Write a paragraph sharing the parenting style your parents used with you.
2. Write a paragraph proving two triangles are congruent.

Use Brainstorming Strategies

What are two questions to think about as you brainstorm?

1.
2.

Lists

Listing involves writing down ____________________ ____________________ and phrase that comes to mind during a set period of time.

To start listing, follow these steps:

1.
2.
3.

Mind Maps

A ____________________ ____________________ helps you see relationships between ideas. To create one, follow these steps:

1.
2.
3.

On Your Own

Now, try mind mapping on your own. Create a mind map on a topic of your choice in the space below.

Think about Audience, Purpose, and Constraints

Define the term **scope.**

What are some questions to ask yourself when considering your audience, purpose, and constraints?

1.
2.
3.

List some common purposes for writing:

Your **audience** is the person or people meant to read or interpret a text.

What questions can you ask yourself as you think about your audience?

1.
    a.
    b.
    c.
2.
    a.
    b.
    c.

On Your Own

Complete the table below, determining the audience for each text.

Text	Audience
Report for work	
Personal blog entry	
An accident report for insurance	
Text message	
Paper for history class	

Lesson Wrap-up

Key Terms

Define the following Key Terms from this lesson.

Audience:

Brainstorming:

Constraint:

Entertaining Text:

Focus:

Informative Text:

Listing:

Mind Mapping:

Narrowed Topic:

Paragraph:

Persuasive Text:

Pre-writing:

Purpose:

Reflective Text:

Scope:

Writing Prompt:

Lesson 4R.3

Writing a Topic Sentence

OBJECTIVES

★ Recognize strong controlling ideas for specific purposes.
★ Recognize the characteristics of a good topic sentence.
★ Understand the differences between a topic and a controlling idea.

BIG IDEA

A clear and concise topic sentence is the foundation for a good paragraph. It guides you and your audience through your writing.

In this lesson, you will learn how to draft an effective topic sentence by reading about the following:

- Characteristics of a Topic Sentence
- Determining a Controlling Idea
- Constructing the Topic Sentence

Characteristics of a Topic Sentence

A topic sentence should be a ______________________________, not a question.

A topic sentence should be _________________ and _________________________.

A topic sentence should be written in your _______________ __________________.

On Your Own

Identify which of the following topic sentences is strongest.

a) College tuition is too high, forcing most students to take on overwhelming student loans.

b) Why is college tuition so high?

c) According to The Money Tracker, college tuition has risen at least 20% a year for the last five years.

Determining a Controlling Idea

Define the term **controlling idea**.

On Your Own

For each purpose, fill in your own controlling ideas about the topic "rising food costs" in the table below.

Purpose	Controlling Idea
To inform	
To persuade	
To reflect	
To entertain	

Constructing the Topic Sentence

________________ + ______________________ ______________ = Topic Sentence

The strongest controlling idea fits all the requirements of a good topic sentence.

Complete the following table by adding a controlling idea and topic sentence for each topic. Fill in a topic of your choice for the last two rows. Remember to determine your purpose.

Topic	Controlling Idea	Topic Sentence
My favorite hobby		
Applying for college		
The effects of social media		

Lesson Wrap-up

Key Terms

Define the following Key Terms from this lesson.

Audience:

Controlling Idea:

Entertaining Text:

Informative Text:

Narrowed Topic:

Persuasive Text:

Purpose:

Reflective Text:

Supporting Detail:

Topic Sentence:

Lesson 4R.4

Organizing a Paragraph

OBJECTIVES

- ★ Identify different types of organizational patterns.
- ★ Recognize appropriate transitions.
- ★ Recognize characteristics of organizational patterns.

BIG IDEA

Organizational patterns help you stay on track while writing and help your audience better understand your ideas.

In this lesson, you will learn the characteristics of six organizational patterns:

- Cause and Effect
- Chronological
- Compare and Contrast
- Order of Importance
- Spatial
- Topical

Cause and Effect

A **cause and effect** paragraph discusses the ________________ and/or ________________ of a topic.

Transitions are ________________, ________________, or ________________ that show ________________________ between ideas.

List three transition words or phrases for a cause and effect paragraph:

1.

2.

3.

Chronological

A **chronological** paragraph shares ideas or events in the ________________ that they occurred.

List three transition words or phrases for a chronological paragraph:

1.

2.

3.

On Your Own

Which of the following passages is written chronologically?

a) Last, going with the natural grain of the wood, apply the stain in an even layer. First, use sandpaper to smooth the wood and remove the existing varnish. Next, clean the surface with mineral spirits.

b) First, use sandpaper to smooth the wood and remove the existing varnish. Last, going with the natural grain of the wood, apply the stain in an even layer. Next, clean the surface with mineral spirits.

c) First, use sandpaper to smooth the wood and remove the existing varnish. Next, clean the surface with mineral spirits. Last, going with the natural grain of the wood, apply the stain in an even layer.

Compare and Contrast

The **compare and contrast** organizational pattern is used for paragraphs that discuss the

______________________________ and ______________________________ between two topics.

List three transition words or phrases for a compare and contrast paragraph:

1.

2.

3.

Order of Importance

A paragraph arranged in **order of importance** organizes ideas from ____________________

______________________________ to ____________________ ______________________________, or vice versa.

List three transition words or phrases for an order of importance paragraph:

1.

2.

3.

On Your Own

Read the following sentences and identify the transition words or phrases.

> The most important step in constructing a well-built house is using quality building materials. For contractors, hiring professional workers is an equally important consideration.

Spatial

A **spatial** paragraph describes a ______________ ____________________ characteristics. The description in a spatially ordered paragraph usually moves from one _________________ to the ____________________ in an ________________________ way.

On Your Own

In the space below, use spatial organization to write a short description of your current location.

Topical

Topical paragraphs are not arranged in a ___________________________ order because all the information is of ________________ ______________________. This pattern is sometimes referred to as "_______________________" and is often used for ____________________ or _____________ writing.

Provide the usual structure of a topical paragraph:

-
-
-
-
-

List three transition words or phrases appropriate for a topical paragraph:

1.
2.
3.

On Your Own

Determine what type of organizational pattern is used in the following paragraph and check the box next to your answer.

> When baking a cherry pie, begin by placing a ready-made pie crust in a nine-inch pie plate. Next, add a can of cherry pie filling to the crust. After spreading out the cherries, place a second crust on the top of the filling. Finally, bake the pie at 400 degrees for about an hour.

- ☐ Cause and Effect
- ☐ Chronological
- ☐ Compare and Contrast
- ☐ Order of Importance

Lesson Wrap-up

Key Terms

Beside each definition, write the corresponding term.

________________________: A word, phrase, or sentence that shows a relationship between ideas

________________________: An organizational pattern used to explain the causes or effects of a topic

________________________: An organizational pattern used to describe a subject's physical characteristics, usually from top to bottom, left to right, or front to back

________________________: An organizational pattern used to show the similarities and differences between two topics

________________________: A text that convinces its audience to adopt a belief or take an action

________________________: A text that gives the audience information about a topic

________________________: The structure of a written text, used to arrange the main points of a work

________________________: A general organizational pattern used for equally important main points

________________________: A text that shares a personal experience or belief

________________________: An organizational pattern that arranges ideas from least to most significant, or vice versa

________________________: A text that explores a topic or event in a creative or humorous way

________________________: An organizational pattern that arranges ideas or events in the order that they occur (also known as sequential order)

Lesson 4R.5

Drafting a Paragraph

OBJECTIVES

- ★ Identify types of transitions and their uses.
- ★ Recognize the eight types of supporting details.
- ★ Recognize the structure of a paragraph.
- ★ Understand the drafting process.

BIG IDEA

Drafting is when you put your ideas together and organize them effectively. The focus should be on building a solid paragraph, not perfecting your grammar and spelling.

In this lesson, you will learn three important elements of a first draft:

- Support Sentences
- Concluding Sentences
- Transitions

Support Sentences

Support sentences can build your ________________________, explain a ____________________ topic, or tell a __________________.

Supporting details are pieces of information that make your paragraph more ________________________ and __________________________ by expanding on your ____________________ ____________________.

List the seven types of supporting details:

1.
2.
3.
4.
5.
6.
7.

On Your Own

Read the following passage and determine the type of supporting detail it uses. Check the box next to your answer.

> It's crucial that states develop better programs for providing children with adequate clothing. When I was ten, my family couldn't afford to buy me winter shoes. Instead, I wore gym shoes as I stood in a foot of snow at the bus stop every day. Eventually, my feet became frostbitten, and I missed school for a week. Worse, I almost lost some of my toes.

- ☐ Expert Analysis
- ☐ Fact
- ☐ Reflection
- ☐ Statistic

Concluding Sentences

Your concluding sentence ends the paragraph by ______________________________ the ideas you just ______________________________.

Often, the concluding sentence also ______________________ the topic sentence using ___________________________ words.

On Your Own

Read the following topic sentence:

> The libraries of today use digital technologies to provide many services that were traditionally completed by hand.

Identify the concluding sentence that best restates the topic sentence.

a. Libraries provide a variety of services.

b. Modern libraries use electronic resources rather than traditional paper and pencils for many of their services.

c. Computer labs can be found at most modern libraries.

Your ______________________ for writing will also influence the _________________ of concluding sentence that fits your __________________________ best.

What are two questions you can ask yourself to think of ideas for your concluding sentence?

1.

2.

Transitions

Define the term **transition.**

Transitions give your audience _____________________________ that they can use to follow the _________________________________ of your __________________.

What are the two main types of transition words?

1.

2.

Complete the following table with the transition words provided in the lesson.

Order	Connections

On Your Own

Read the following sentences and identify the one that uses a stronger transition word.

a. I need to deposit my check; likewise, the bank is closed.
b. I need to deposit my check; however, the bank is closed.

Lesson Wrap-up

Key Terms

Define the following Key Terms from this lesson.

Audience:

Concluding Sentence:

Drafting:

First Draft:

Informative Text:

Main Idea:

Organizational Pattern:

Persuasive Text:

Purpose:

Reflection:

Support Sentence:

Supporting Detail:

Topic Sentence:

Transition:

Chapter 5R
Integrated Review: Reading Critically

Lesson 5R.1

Pre-Reading Strategies

OBJECTIVE

★ Learn basic pre-reading strategies.

BIG IDEA

Pre-reading activities prepare your mind to get the most out of the time you spend studying and reading. In this lesson, you will learn three important strategies for pre-reading:

- Scan the Title, Table of Contents, and Major Headings
- Make Predictions and Find Connections
- Create a Plan to Start Reading

Scan the Title, Table of Contents, and Major Headings

Title

Before you begin a reading, look for a ______________________ about its contents from the title.

On Your Own

Read the following titles. What would you expect to read about in each of the texts?

1. "Shortcut to Victory: Inside the Iowa Primary"
2. *Nickel and Dimed*
3. "Opposition to the Vietnam War and the Effect of the Arts"
4. "College Athletics: Perspectives on the Winners and the Losers"
5. *The Bluest Eye*

Table of Contents

The **table of contents** shows you how the information in a book or long document is

___________________________.

Headings and Topic Sentences

Bold _________________________ and _____________________ will give you clues about what each section in the reading covers.

Topic sentences usually appear at the ______________________________ of a **paragraph**, and if they are well-written, they should give you a strong idea of what the paragraph will ________________________.

On Your Own

Consider the following topic sentences. What would you expect to read in the paragraph that each sentence introduces?

> Social media is a great way to keep in touch with family and friends.
>
> The researchers soon realized that their original study had major issues.
>
> Despite this opposition, I would argue that the benefits of this business model will outweigh the costs.

Make Predictions and Find Connections

Making predictions will prepare you to think more ____________________________ about a reading.

List the six questions you can ask yourself when making predictions:

1.
2.
3.
4.
5.
6.

READING APPLICATION

Before reading *Taking Care of Beth*, preview its contents and consider how you can make predictions and connections to better understand its meaning. Then, answer the following questions.

What does the title suggest about the story?

Based on the beginning of the story, what can you already tell about the main characters?

What does this tell you about Irving's past?

What is the author trying to tell you about the relationship between Beth and Irving?

What does the author want you to gain from this story?

Define the term **prior knowledge**.

A **KWL chart** is a useful tool in making ______________________________ between your prior knowledge and your study material.

What do the letters stand for?

K ____

W____

L ____

READING APPLICATION

Before reading *Ghost in the Machine: Debunking the Paranormal*, apply pre-reading strategies to interact with the text. Utilize the blank KWL chart provided below to guide your pre-reading process, and return to it after reading to note what you learned.

Ghost in the Machine: Debunking the Paranormal

What I Know	What I Want to Know	What I Learned

Create a Plan to Start Reading

List the three questions you can ask yourself as you develop your reading plan:

1.

2.

3.

Lesson Wrap-up

READING APPLICATION

Read the passage *Considerations for Future Dog Owners* and use the annotations as a pre-reading guide.

What is the main idea of each paragraph?

Paragraph 1:

Paragraph 2:

Paragraph 3:

Paragraph 4:

Paragraph 5:

Paragraph 6:

Key Terms

Define the following Key Terms from this lesson.

Abstract:

Keywords:

KWL Chart:

Paragraph:

Pre-reading:

Prior Knowledge:

Table of Contents:

Topic Sentence:

Lesson 5R.2

Finding Meaning Through Visuals

OBJECTIVES

★ Recognize different types of charts and graphs.
★ Understand how visual clues add to the meaning of a text.

BIG IDEA

Visual clues such as charts, graphs, pictures, icons, colors, and shapes can add just as much meaning to a text as the actual words do. You need to analyze these visuals carefully so that you don't miss valuable information.

This lesson will help you identify and understand three types of visual clues:

- Font Design
- Images
- Charts and Graphs

Font Design

Reading through the titles and headings will help you get a better ______________________________ of your ______________________________.

On Your Own

You've been invited to a formal wedding. Which font do you think will appear on the invitation?

You're Invited *You're Invited* You're Invited

READING APPLICATION

Review Tabitha Hawkins' résumé, paying attention to the use of font, italics, spacing, and lines. Below, take notes on how these decisions impact the overall effectiveness of the résumé.

Font:

Italics:

Spacing:

Lines:

Images

What is the first step of analyzing an image?

What four questions should you ask yourself when reflecting on your prior knowledge?

1.
2.
3.
4.

Charts and Graphs

Structure

Bar Graphs

Bar graphs are used to ______________________ and ____________________ groups of information.

Pie Charts

Pie charts show ______________________________ of a whole.

Line graphs

Line graphs show ____________________ over __________________.

On Your Own

Which visuals would fit the following sets of data best? Complete the chart below.

Data	Visual
A comparison of sales revenue by restaurant location	
The growing number of Pinterest users since 2010	
The heights of your family members	
A comparison of average grocery costs over the last twelve months	
The average cost of rent in three different cities	

Data

The __________________ will give you a clear idea of the data being illustrated in the chart or graph.

You can then read the **legend**, which shows you the ___________________ or ____________________ that represent each group of information.

Finally, read the ______________________ and ______________________ on each part of the chart or graph. This will show you how the data is ________________________ together.

Lesson Wrap-up

READING APPLICATION

Read the passage about minimum wage and answer the following questions.

What is the overall topic of this passage?

What terms might you need to define during your pre-reading activities?

What kind of visual is used in the passage?

Key Terms

Define the following Key Terms from this lesson.

Bar Graph:

Data:

Font:

Legend:

Line Graph:

Pie Chart:

Visual Clue:

Lesson 5R.3

Active Reading Strategies

OBJECTIVES

★ Learn a process for active reading.
★ Understand the differences between active and passive reading.

BIG IDEA

Active reading is a process that focuses your attention on what you are reading so that you can understand and retain the material.

Define the term **passive reading**.

This lesson will teach you how to use active reading strategies to get the most from your reading.

- Make Pre-Reading a Priority
- Check Your Understanding
- Review What You Learned

Make Pre-Reading a Priority

What steps should you follow during pre-reading?

1.

2.

3.

4.

READING APPLICATION

Read the passage *Rules for Celebrity Ads on Social Media.* As you read, answer the annotation questions below.

How can you connect to this idea? Where have you seen celebrities in advertising lately?

Now that you've finished reading, consider how this information built upon what you already knew.

Check Your Understanding

Summarize Paragraphs

Define the term **summary**.

Summarizing each paragraph will help you identify areas where you did not ____________________ the reading.

READING APPLICATION

Practice checking your understanding as you read the passage about attraction.

Review What You Learned

The final stage in the active reading process involves four steps:

1.
2.
3.
4.

Lesson Wrap-up

READING APPLICATION

Read the passage *Battle of the Parties: Supreme Court Edition*. As you read the passage, consider how you can apply active reading strategies to comprehend the author's meaning.

What is the topic of the passage?

Summarize the author's message.

Test Yourself

Next to each question, write **Y** for Yes or **N** for No. Check your work using the Answer Key in the back of the book.

1. __________ Sarah is reading the chapter as quickly as possible and looks at Facebook every five minutes to see if anyone commented on her photo. Is she practicing active reading?
2. __________ Jose is doing a KWL chart and looking up unfamiliar words. He also looked at all the bold headings in his reading material. Is he practicing active reading?

3. ____________ Jane has been assigned a chapter on communism. She remembers something she learned about the Cold War that is related to this subject. She is looking up that information and thinking about what she knows. Is she activating her prior knowledge?

Key Terms

Define the following Key Terms from this lesson.

Active Reading:

Conclusion:

Introduction:

KWL Chart:

Passive Reading:

Reading Comprehension:

Summary:

Lesson 5R.4

Finding Connections and Patterns

OBJECTIVES

★ Recognize patterns in the vocabulary and details an author uses.
★ Understand and apply strategies for thinking critically about a reading.

BIG IDEA

You should draw connections between ideas and details within a text to understand its meaning on a deeper level. Additionally, identifying patterns within a text can reveal an author's purpose for writing.

In this lesson, you will learn strategies for finding connections and patterns in your reading:

- Outline the General and Specific Information
- Locate Patterns in Vocabulary
- Locate Patterns in the Details

Outline the General and Specific Information

Texts include a combination of related general and specific information.

Add your own example of general and specific information in the space provided.

General	Specific
snacks	plantain, ramen, macaroni and cheese

An outline is a helpful way of reviewing the general and specific information in a text so you can better understand its purpose.

What questions can outlining help answer about a text?

READING APPLICATION

Read the passage *How to Keep Texting Language from Affecting Our Communication*. Read it a second time, and write an outline to sort the information in each paragraph. Then, answer the following questions:

How has the author arranged ideas and details in the text?

How do the details relate to the topic and main idea of the text? Is the relationship clear?

In your own words, write a summary of the passage's general information and the specific solutions to improve the issue.

Locate Patterns in Vocabulary

What are strategies you can use for locating patterns in vocabulary?

What can patterns in vocabulary reveal about the author's intentions?

Locate Patterns in the Details

What are strategies you can use for locating patterns in the details an author uses?

What can patterns in the details reveal about the author's intentions?

Lesson Wrap-up

Key Terms

Define the following Key Terms from this lesson.

Major Detail:

Minor Detail:

Paragraph:

Purpose:

Supporting Detail:

Topic Sentence:

Lesson 5R.5

Finding Meaning Through Context

OBJECTIVE

★ Recognize the different types of context clues.

BIG IDEA

Context clues will help you discover the meanings of unfamiliar words through clues found in the text. You don't always have access to Google or the dictionary, so learning to use context clues will help you better understand the reading material.

Define the term **context clue**.

In this lesson, you will learn how to find the meaning of a word based on context clues:

- Look for Clues in Nearby Sentences
- Use Substitution to Test Possible Meanings

Look for Clues in Nearby Sentences

What is the first step in finding the meaning of an unfamiliar word?

Definition

Example:

> Imagine my humiliation when my friends tricked me into wearing a silly costume to the fancy dinner party. I've never been that embarrassed before.
>
> Based on the context clue, what is the meaning of *humiliation*?

Synonym

A **synonym** is a word that has the ____________________ meaning as another word. Authors sometimes use a ____________________ and a ___________________________ to add extra emphasis or explanation to the sentence.

Example:

> The spoiled meat was so putrid, or smelly, that I had to plug my nose.
>
> In this sentence, the meaning of the word *putrid* is the same as the word *smelly*.

In the following sentence, choose the synonym context clue for the word *obedient*.

Even though my dog, Pippa, lacks formal training, she is quite obedient, or well-behaved, by nature.

Antonym

An **antonym** is a word that has the ______________________________ meaning of another word.

List three signal words that indicate the use of an antonym:

1.

2.

3.

In the following sentence, choose the antonym context clue for the word *scorching*.

I expected the temperatures here to be scorching, but they're actually quite cool.

Based on the context clue, what is the meaning of *scorching*?

Example

Authors sometimes use ______________________________ to explain the meaning of a difficult word.

In the following sentence, indicate the example that explains the meaning of *detests*.

Sarah detests most kinds of seafood; for example, she can't stand the taste of lobster.

Inference

An **inference** is a ____________________________ _________________________________ based on what's happening in the text.

Read the following example and use inference to explain the meaning of *gridlock*.

Every morning during rush hour, the gridlock on the freeway is so bad that it takes me an hour to get to work.

Use Substitution to Test Possible Meanings

To use substitution, think of a possible _________________________ and read the sentence with that word.

Example:

Each time it snows, I feel the overwhelming need to hibernate; I rarely leave my house, and it takes me forever to get anything done.

Use substitution to explain the meaning of *hibernate*.

On Your Own

Read the following paragraph and use context clues to determine the meanings of the underlined words in the space below.

> Nutritional science is effectively a 20th century creation. On the basics of what the human animal ought to eat, as expressed in dietary recommendations around the world, the experts more or less agree (Cannon). However, on many specific issues there are differences of view that are intense and even polemical in character. Obesity is one area of conspicuous contention. In part, these disputes reflect the youth of the science. They also reflect the practical difficulties of conducting long-term, controlled research on the daily routines of numerous human subjects. The consequence is that many non-nutritionists working on obesity have difficulty finding agreed, actionable conclusions from nutritional science on what to do about the problem.

dietary:

polemical:

contention:

actionable:

Lesson Wrap-up

Test Yourself

Answer the following questions. Check your work using the Answer Key in the back of the book.

1. What is an antonym for *happy*?

2. What kind of context clue is being used in the following sentence?

   The once feral cat became tame once she got to know us.

3. What is a synonym for *frightened*?

4. What kind of context clue is being used in the following sentence?

   The elderly man was not stable after his surgery. He fell down multiple times.

5. What is the signal word indicating the use of an antonym in the following sentence?

   The moon was radiant tonight, unlike last night when it was hidden behind the clouds.

Key Terms

Define the following Key Terms from this lesson.

Antonym:

Context Clue:

Example:

Inference:

Paragraph:

Synonym:

Lesson 5R.6

Finding Meaning Through Word Parts

OBJECTIVE

★ Identify the meaning of a word from word parts.

BIG IDEA

You can use word parts to piece together the meanings of unfamiliar words.

This lesson will teach you how to use word parts to find the meanings of words.

Roots

The ____________________ is the main part of a word and contains the basic ____________________ of the word. ________________ ____________________ are roots that can stand alone without any other word parts.

________________________ ____________________ form the ________________ of a word but cannot stand __________ their _____________.

Describe what you think each of the following words means based on the shared root, *auto*.

autobiography:

automatically:

autonomy:

Define the term **word family**.

Prefixes

Prefixes are a second type of word part that are added to the ________________________ of roots to create new words.

To complete the following table from the lesson, add the meaning of each common prefix.

Prefix	Meaning
bi-	
co-	
de-	
mis-	

non-, un-	
post-	
pre-	
re-	
semi-	
trans-	
uni-	

Suffixes

Suffixes are added ____________________ a ____________________ to change the meaning of a word.

To complete the following table from the lesson, add the meaning of each common suffix.

Suffix	Meaning
-able	
-ance, -ence	
-ful	
-fy	
-ion, -sion, -tion	
-ize	
-less	
-ly	
-ness	
-ology	
-or, -er	

On Your Own

Look at the following words and use the roots, prefixes, and suffixes from this lesson to determine their meanings and complete the following chart.

Word	Meaning
judicator	
graphology	
autophobia	
diction	

Lesson Wrap-up

Key Terms

Define the following Key Terms from this lesson.

Combining Root:

Context Clue:

Prefix:

Root:

Root Word:

Suffix:

Word Family:

Lesson 5R.7

Finding Meaning Through Inference

OBJECTIVE

★ Learn strategies for making inferences about a text.

BIG IDEA

Authors do not always directly state everything they want their readers to know about a topic. It's sometimes up to the reader to make a logical conclusion in order to gain full understanding of the text.

Define the term **inference**.

List the three factors that inferences are based on:

1.

2.

3.

What are some questions you can answer by making inferences?

1.

2.

3.

4.

5.

In this lesson, you will learn how to make inferences about a text using three steps:

- Activate Your Prior Knowledge
- Look for Clues
- Ask Yourself Questions

Activate Your Prior Knowledge

Define the term **prior knowledge**.

Look for Clues

Define the term **implied information**.

Visuals

The ____________________ inside a text can help you understand the author's ______________________ and _________________. Visuals can also help you make connections to your ________________ __________________________.

Tone

Define the term **tone.**

Write one example word for each connotation in the table below.

Connotation	Example
Positive	
Negative	
Neutral	

Figurative Language

Define the following types of figurative language:

Metaphor:

Simile:

Personification:

Facts and Opinions

In the following table, compare and contrast facts and opinions. Include an example of each.

Facts	Opinions

Organization

The **organizational pattern** of a text can contain clues about the author's ____________________ and ____________________ ____________________. As you read, look for ____________________ ____________________ that might indicate the type of organization.

READING APPLICATION

Read the article *First Impressions* and notice how you can infer information about the author's message because of the deliberate tone and organization.

What is the author's purpose of this article as stated in the last sentence of the first paragraph?

Would you say the overall tone of this passage is positive or negative? Explain your answer.

Ask Yourself Questions

To make logical inferences based on the textual clues you've found, ask yourself these questions:

1.
2.
3.
4.
5.
6.

Lesson Wrap-up

READING APPLICATION

Read the narrative excerpt *Uncle Ted* and see what you can infer based upon the author's suggestions. Then, answer the following questions.

What information indicates that Uncle Ted is a farmer?

How do you know Uncle Ted likes the person who is telling the story?

Is Uncle Ted a good cook? How do you know?

Key Terms

Define the following Key Terms from this lesson.

Cause and Effect Organization:

Chronological Organization:

Compare and Contrast Organization:

Connotative Meaning:

Fact:

Figurative Language:

Inference:

KWL Chart:

Main Idea:

Metaphor:

Opinion:

Order of Importance Organization:

Organizational Pattern:

Personification:

Pre-reading:

Prior Knowledge:

Purpose:

Reading Comprehension:

Simile:

Spatial Organization:

Tone:

Topical Organization:

Lesson 5R.8

Types of Main Ideas and Evidence

OBJECTIVES

★ Identify different types of supporting details.
★ Recognize the main idea in a text.

BIG IDEA

Identifying main ideas and supporting details in a text can help you understand it more thoroughly. Also, being able to pinpoint the types and locations of supporting evidence allows you to decide whether or not the evidence effectively supports the main idea.

In this lesson, you will learn about the following:

- Recognizing the Main Idea(s)
- Recognizing Types of Evidence
- Analyzing the Supporting Details

Recognizing the Main Idea(s)

A topic is the most ______________________ characteristic of a ________________. The main idea is the more ______________________ claim the author makes *about* the ________________.

READING APPLICATION

Read the excerpt from the essay on the evolution of technology.

In your own words, what is the paragraph's main idea?

Implied main ideas are often used when the author is writing to ____________________ or ______________________.

Thesis and Purpose Statements

In a ______________________ text, the main ideas of each ____________________________ serve to make claims that support a bigger idea.

Define the term **thesis statement**.

Define the term **purpose statement**.

Where are purpose statements most commonly found?

Choose an example thesis and purpose statement provided in the text and explain what evidence might support the statement.

Thesis Statement

Example:

Possible supporting evidence:

Purpose Statement

Example:

Possible supporting evidence:

Locating the Thesis

Try following these steps to locate the thesis statement:

1. Find the ____________ ____________ of each ____________.
2. Form the ____________ ____________ into a ____________ paragraph.
3. Determine the ____________ ____________ of the summary paragraph to find the ____________.

Recognizing Types of Evidence

____________ ____________ are the specific pieces of information that are used to support a main idea.

Supporting details often answer questions that start with ________, ________, ________, ________, ________, and ________.

There are several types of information that an author can use to support main ideas. Define each term:

Anecdotes:

Descriptions:

Examples:

Expert Analysis:

Facts:

Reflections:

Statistics:

Analyzing the Supporting Details

The ____________________ and ________________________ of the text will influence the types of supporting details that the author uses.

List the questions you can use to determine how well supporting details support a main idea:

1.

2.

3.

Read the paragraph and answer the following questions.

> Every year, World Malaria Day forces us to look at where we came from, where we are, and what still needs to be done. Joint action over the past decade has led to an impressive impact: malaria infection rates have been cut in half, and 4.3 million lives have been saved ("World Malaria Report 2014"). Fifty-five countries are on track to reach the World Health Assembly target of a 75% reduction in their malaria burden by 2015 ("World Malaria Report 2014"). Although these huge gains are impressive, they remain fragile if the momentum of the joint action cannot be maintained. Clearly, not keeping the momentum leads to the resurgence of malaria, as we have experienced in numerous previous elimination efforts at national or subnational level.

What is the main idea of the paragraph?

What two statistics does the author use to support the main idea?

1.

2.

Lesson Wrap-up

READING APPLICATION

Read the persuasive essay titled *Arts Programs*. As you read, take note of the author's claims and supporting details. Then, answer the questions below.

What is the thesis statement?

What evidence is used to support the thesis statement? Provide two examples.

1.

2.

Key Terms

Define the following Key Terms from this lesson.

Description:

Evidence:

Implied Main Idea:

Introduction:

Main Idea:

Purpose:

Purpose Statement:

Signal/Signpost Word:

Thesis Statement:

Lesson 5R.9

Identifying Purpose and Tone

OBJECTIVES

★ Recognize the purpose of a text.
★ Understand how tone adds to the meaning of a text.

BIG IDEA

Authors use purpose and tone as tools to help their readers better understand what they are communicating. As a reader, you must think critically about the *why* and the *how* behind a text. This will give you a more complete understanding of what an author is trying to say.

In this lesson, you will learn how to identify the following:

- Author's Purpose
- Author's Tone

Author's Purpose

To Inform

The author of an **informative** text wants to give the audience ______________________ about a topic.

Name three texts that commonly seek to inform:

1. 2. 3.

To Persuade

The author of a **persuasive** text wants to ____________________ the audience to adopt a

_________________ or take an ____________________.

Name three texts that commonly seek to persuade:

1. 2. 3.

To Narrate

The author of a **narrative** text wants to ________________ and ________________ on a personal

______________________ or ______________________.

Name three kinds of texts that commonly seek to narrate:

1. 2. 3.

To Entertain

The author of an **entertaining** text wants to ____________________ the reader by ____________________ a topic or event in a ____________________ or ____________________ way.

Name three kinds of texts that commonly seek to entertain:

1. 2. 3.

Author's Tone

Tone is the ____________________ expressed about the **topic** of the text. Authors use ______________ and ___________________ carefully to establish a clear tone. Tone usually falls under the following broad categories: ________________, _________________________, or _________________________.

Complete the table using adjectives that describe the tone of a text. There are some ideas in the lesson, but try to think of your own as well.

Positive	Negative	Neutral

In most cases, tone is not stated _____________________; instead, the reader has to ________________ __________________ ________ __________________ to determine how the author feels about the topic.

Look at the following examples and consider how word choice affects the overall tone of the sentence.

My dad bought my mom an affordable diamond ring for their anniversary.

My dad bought my mom a cheap diamond ring for their anniversary.

For each word in the table, think of a synonym with the opposite tone. Look up any unfamiliar words.

Negative	Positive
	thrifty
	outgoing
reeks	
cowardly	

List four questions you can ask yourself to determine an author's tone:

1. 3.

2. 4.

Indicate the positive language in the following sentences.

> After the tornado swept through the town, the community immediately supported everyone who had suffered losses. Friends, neighbors, and strangers united in a way that transformed the entire area.

Lesson Wrap-up

Test Yourself

Next to each question, write **E** for entertain, **I** for inform, **P** for persuade, or **R** for reflect. Check your work using the Answer Key in the back of the book.

1. ______ What is most likely the purpose of a comic book?
2. ______ What is most likely the purpose of a science textbook?
3. ______ What is most likely the purpose of a children's fiction book about the life of a unicorn?
4. ______ If an author says that childhood vaccines should be mandatory, what is his or her purpose?
5. ______ If an author writes about childhood experiences in Lebanon, what is his or her purpose?
6. ______ If an author tries to be as objective and unbiased as possible, what is his or her purpose?

Next to each statement, write **T** for True or **F** for False. Check your work using the Answer Key in the back of the book.

7. _______ To understand a text fully, it's important to know the author's purpose.
8. _______ An author would never be dishonest about his or her true purpose for writing.
9. _______ A text with a neutral tone just gives the facts in a straightforward way.
10. _______ A textbook will generally have a neutral tone.

Key Terms

Write the letter of each term next to its definition.

a. Tone
b. Persuasive Text
c. Informative Text
d. Entertaining Text
e. Reflective Text

______ A text that explores a topic or event in a creative or humorous way

______ A text that gives the audience information about a topic

______ A text that convinces its audience to adopt a belief or take an action

______ A text that shares a personal experience or belief

______ The positive, negative, or neutral attitude that an author expresses about a topic

Chapter 6R
Integrated Review: Writing Critically

Lesson 6R.1
Determining Essay Focus

OBJECTIVES

★ Identify strategies for choosing a topic and narrowing scope.
★ Understand a variety of purposes when writing a longer text.
★ Understand the importance of narrowing the scope of a topic.

BIG IDEA

The scope, or focus, of your topic should be appropriate for the length of your writing. The scope you choose for your topic should be thoroughly addressed in the given length for the assignment.

Define the term **scope.**

In this lesson, you will learn three useful steps in choosing a topic and narrowing your scope:

- Review Instructor Guidelines
- Use Brainstorming Strategies
- Think about Purpose, Audience, and Constraints

Review Instructor Guidelines

What are the six questions to ask yourself when you are trying to come up with ideas for a topic?

1.
2.
3.
4.
5.
6.

Use Brainstorming Strategies

Free-writing

List the steps of free-writing:

1.

2.

3.

On Your Own

Practice free-writing using the following space. Add your topic at the top; then, start a timer. Keep writing down ideas until the timer goes off.

Grouping

Define the term **grouping.**

List the three steps of grouping:

1.

2.

3.

Think about Audience, Purpose, and Constraints

List some common purposes for writing:

Define the term **audience**.

List some constraints your professor or manager might give you when you have a written assignment:

1.
2.
3.

Lesson Wrap-up

Key Terms

Beside each definition, write the corresponding term.

______________________:	The style and arrangement of a text
______________________:	Exploring and developing ideas
______________________:	A limitation that affects your writing
______________________:	The people who read your writing
______________________:	Clear communication and support of a main idea
______________________:	A brainstorming strategy for getting your ideas down on paper
______________________:	A stage of writing that involves making decisions, planning ideas, and identifying assignment guidelines
______________________:	A topic that you have made more specific by considering your purpose, audience, and constraints
______________________:	A brainstorming strategy that helps you organize your thoughts
______________________:	A short piece of writing that focuses on one main idea
______________________:	The goal of a text
______________________:	The focus of a writing assignment
______________________:	What you already know about a topic

Lesson 6R.2

Writing a Thesis or Purpose Statement

OBJECTIVES

★ Recognize the characteristics of a good thesis statement.
★ Understand the difference between a thesis statement and a statement of purpose.

BIG IDEA

Readers will know what to expect from your writing when you construct a well-written thesis statement or purpose statement. This statement will also give you a clear structure for your paper.

This lesson will discuss three steps in creating a thesis statement:

- Decide Between a Thesis and Purpose Statement
- Combine a Narrowed Topic and Controlling Idea
- Strengthen Your Thesis Statement

Decide Between a Thesis and Purpose Statement

Define the term **thesis statement**.

In what type of writing are thesis statements especially common?

Where do thesis statements almost always appear?

What do **purpose statements** announce and preview?

In what kind of reports are purpose statements most commonly found?

Where do purpose statements appear?

Combine a Narrowed Topic and Controlling Idea

The first step in creating a thesis statement is ______________________ ______________

______________ so that you can choose a ______________________ ______________.

Define the term **controlling idea**.

On Your Own

Read the following controlling ideas for the topic "student loan debt" and determine the purpose. Check the box next to your answer.

Student loan debt should be regulated by the government.

- ☐ Persuade
- ☐ Inform

Student loan debt is the subject of numerous proposed government reforms.

- ☐ Persuade
- ☐ Inform

On Your Own

Practice creating your own thesis statements by combining narrowed topics and controlling ideas in the text boxes below.

Narrowed Topic	+	Controlling Idea
	+	
	+	

Strengthen Your Thesis Statement

One way to strengthen a thesis would be to include the ______________________________ for your ______________________________. To create this type of thesis, add a **subordinating conjunction** like *because* or *including*; then, insert a brief _________________________ of your reasoning.

Example:

Lesson Wrap-up

Key Terms

Define the following Key Terms from this lesson.

Argument:

Audience:

Controlling Idea:

Drafting:

Hedging Word:

Informative Text:

Introduction:

Main Idea:

Narrowed Topic:

Paragraph:

Persuasive Text:

Pre-writing:

Purpose:

Purpose Statement:

Revising:

Subordinating Conjunction:

Thesis Statement:

Topic Sentence:

Lesson 6R.3

Organizing and Outlining an Argument

OBJECTIVE

★ Identify the characteristics and purposes of different types of outlines.

BIG IDEA

Organizing your writing will keep you from getting lost during the writing process and will keep your writing clear and easy to follow.

In this lesson, you will learn how to use three different types of outlines:

- Working Outline
- Topic Outline
- Sentence Outline

Define the term **outline.**

Working Outline

Working outlines are ______________________________________.

List the two important functions of a working outline:

1.

2.

What should you do if you find yourself moving in a slightly different direction than you originally planned?

Topic Outline

Topic outlines use short ________________ to represent the ___________________________ of your paper.

Organizational Patterns

List six organizational patterns:

1.

2.

3.

4.

5.

6.

Building the Outline

Fill in the topic outline below with your own ideas about adopting pets versus purchasing them.

I. Adopting pets

  A.

    1.

    2.

  B.

    1.

    2.

II. Purchasing pets

  A.

    1.

    2.

  B.

    1.

    2.

Sentence Outline

What are the similarities and differences between sentence outlines and topic outlines?

Lesson Wrap-up

Key Terms

Define the following Key Terms from this lesson.

Heading:

Organizational Pattern:

Outline:

Sentence Outline:

Subheadings:

Topic Outline:

Working Outline:

Lesson 6R.4

Combining Words or Sentences

OBJECTIVES

- ★ Understand how to combine words effectively.
- ★ Understand how to combine sentences effectively.

BIG IDEA

You can combine words and sentences to communicate your ideas effectively and make your writing easy to understand.

In this lesson, you will learn the following concepts:

- How to Combine Words
- How to Combine Sentences

How to Combine Words

What types of words can be combined?

Coordinating conjunctions join similar ______________ or ______________ of words together.

In the following space, use coordinating conjunctions to write three examples of combined words.

1.

2.

3.

Correlative conjunctions are similar to ______________ ______________; they both connect two or more word groups of ______________ ______________.

How to Combine Sentences

An **independent clause** can stand alone as a ______________ ______________. A **dependent clause** is a word group containing a ______________ and a ______________ that cannot stand alone as a complete sentence.

List the combinations in each sentence pattern:

Simple Sentence	
Compound Sentence	

Complex Sentence	
Compound-Complex Sentence	

What methods can be used to combine clauses?

Lesson Wrap-up

Key Terms

Define the Key Terms from this lesson.

Adjective:

Adverb:

Clause:

Complex Sentence:

Compound Sentence:

Compound-complex Sentence:

Conjunction:

Coordinating Conjunction:

Correlative Conjunctions:

Dependent Clause:

FANBOYS:

Independent Clause:

Noun:

Verb:

Lesson 6R.5

Emphasizing Words or Phrases

OBJECTIVE

★ Learn how to place emphasis on words or phrases in a sentence.

BIG IDEA

You can use words and structures to emphasize the most important details in your writing.

In this lesson, you will learn three ways to emphasize important words or phrases:

- Subordination
- Word Order
- Sentence Structure

Subordination

Usually, the main **subject** and **verb** are the most ________________________ words in a sentence. Less important ideas should be **subordinated**, or ________________________, in **dependent clauses.**

On Your Own

Read the following passage and identify the sentence that uses subordination.

> The famous pirate Blackbeard blockaded the port of Charleston, South Carolina, in 1718. He demanded medical supplies in exchange for the safe return of a group of prisoners. After the exchange was delayed, the pirates moved several ships into the harbor. The city was alarmed by this; Blackbeard's reputation made every coastal city fear him.

Word Order

You can make important information more ________________________ by placing it at the beginning or end of a sentence. Similarly, you can emphasize **phrases** by moving them to the beginning of a sentence. By ____________________ the subject and verb of the sentence, you can ____________________ a feeling of suspense in your audience.

Sentence Structure

The third way to emphasize ideas is by using a ________________________ or ____________________________ sentence structure. These sentences __________________ ______________ to the audience because they are different from the rest of the text.

Test Yourself

Identify the most important idea in each of the following sentences. Check your work using the Answer Key in the back of the book.

1. Even though you have been busy, you still should've called me.
2. Twirling gracefully in the fall breeze, the leaves reminded me of little ballerinas.
3. After careful consideration, I have decided not to take the job in London.
4. Whenever my mother-in-law comes around, I feel like I have been judged on my child-rearing and my housekeeping.
5. In order to be considered for this promotion, employees must submit their Request of Interest form to Human Resources.

Lesson Wrap-up

Key Terms

Write the letter for each term next to its definition.

a) Dependent Clause
b) Independent Clause
c) Phrase
d) Subject
e) Subordinating Conjunction
f) Subordination
g) Verb

______ A word group that contains a subject and a verb and expresses a complete thought

______ The person, place, thing, event, or idea a sentence is about

______ A stylistic method for de-emphasizing an idea in a combined sentence

______ A word that represents an action, relationship, or state of being

______ A word group that contains a subject and a verb but does not express a complete thought

______ A word group that adds meaning to a sentence but does not express a complete thought and usually lacks a subject and a verb

______ A conjunction that introduces a dependent clause

Lesson 6R.6

Using Inclusive Language

OBJECTIVES

★ Identify exclusive language.
★ Understand the importance of inclusive language.

BIG IDEA

By becoming familiar with the differences between inclusive and exclusive language, you can ensure that your writing is respectful and sensitive to others.

In this lesson, you will learn to use language that is inclusive based on the following factors:

- Gender
- Ethnicity or Culture
- Physical or Mental Ability
- Sexual Orientation

Gender

In the past, society has used **gender-specific** words like *mankind* to describe both ________________ and ________________________. In these instances, it would be better to use a **gender-neutral** term like ________________________.

Instead of defaulting to a certain gender, use the term ______________________________.

You can also make the ___________________ ___________________ so that you can use the plural, gender-neutral pronoun *they*.

Add the correct, gender-neutral personal pronoun to the following sentences.

> If someone needs to leave the office early today, __________ or __________ will need to talk to the vice president first.
>
> If the managers need to leave the office early today, __________ will need to talk to the vice president first.

Ethnicity or Culture

Inclusive language does not make ______________________________ about people based on their ______________________ or ______________________.

When you do need to refer to a person's ______________, make sure you use the correct term. If you are unsure, ____________________ _______________ _____________ ______________________________.

Physical or Mental Ability

Just like gender or ethnicity, a person's physical or mental abilities should only be ______________________________ if this information is ____________________________ to your purpose.

Sexual Orientation

If a person's sexual orientation is irrelevant to the sentence, don't ________________________________ it. When you do need to reference someone's sexual orientation, always use terms that are __________________________________.

Lesson Wrap-up

Key Terms

Define the following Key Terms from this lesson.

Credibility:

Exclusive Language:

Gender-Neutral:

Gender-Specific:

Inclusive Language:

Paragraph:

Personal Pronoun:

Pronoun:

Chapter 7R
Integrated Review: Research

Lesson 7R.1
Introduction to Research

OBJECTIVES

★ Understand the basics of research.

BIG IDEA

Knowing the basics of research can help you avoid plagiarism and use reliable sources to support your goals.

In this lesson, you will learn about the following:

- The Basics of Research
- The Basics of Plagiarism
- The Basics of Research Styles

The Basics of Research

What is research?

When is research necessary for writing?

What is a source?

Acknowledging ____________________ and ____________________ them correctly both in your text and in a ____________________ ____________________ list is a basic requirement for academic writing.

The Basics of Plagiarism

Define the term **plagiarism**.

List three examples of intentional plagiarism:

-
-
-

List three examples of unintentional plagiarism:

-
-
-

To avoid accidental plagiarism, use a ____________________ or _______________ _______________ to keep track of your research.

The Basics of Research Styles

MLA

What does MLA stand for?

APA

What does APA stand for?

CMS

What does CMS stand for?

CSE

What does CSE stand for?

Lesson Wrap-up

Key Terms

Beside each definition, write the corresponding term.

______________________: The people who read your writing

______________________: The act of using borrowed words or ideas without giving credit to the author

______________________: A more formal style of writing used to analyze a topic

______________________: An original document or first-hand account that a writer consults for research

Lesson 7R.2

Tools for Purposeful Research

OBJECTIVES

★ Learn how to keep research records.
★ Recognize types of research.
★ Understand the importance of a research timeline.
★ Understand when to use different research methods.

BIG IDEA

Your research should play a meaningful role in the content and organization of your paper. The research process will be more effective if you use the proper tools.

In this lesson, you will learn about three types of research tools:

- Research Methods
- Research Records
- Research Timelines

Research Methods

Library Catalogs and Databases

Academic journals contain ____________________ written by ____________________ or other ____________________ in a particular field.

Most libraries have an ____________________ ____________________ that you can use to search available materials.

Internet Searches

On the internet, you can find ____________________, ________________ ____________________, ________________, and ____________________.

Conducting online research is an excellent way to find source material on ____________________ ________________________ topics like ____________________________ and ____________ ________________.

Field Research

Field research involves personally collecting information through ____________________________ and ________________________.

It's important to consider your assignment ____________________________ just in case you need to ________________________ extra time to gather your research.

Research Records

Research Notes

Define the term **research notes**.

Checklist: Research Notes

✓ ______________________________

✓ ______________________________

✓ ______________________________

✓ ______________________________

Research Journal

A **research journal** is a ____________ of your overall ______________ ______________. Writing down your ________________ as you work helps you keep track of good ____________ or ________________.

Working Bibliography

Define the term **working bibliography**.

Schedule a Research Timeline

Add scheduled ________________ ______________ to your **planner** so that you don't __________________ or fall behind. Add any research ________________ that your instructor has required.

List two examples of special situations that may require extra planning:

1.

2.

Lesson Wrap-up

Key Terms

Define the following Key Terms from this lesson.

Academic Journal:

Credibility:

Field Research:

Planner:

Pre-writing:

Research Journal:

Research Notes:

Working Bibliography:

Writing Center:

Chapter 8R
Integrated Review: Unique Forms of Writing

Lesson 8R.1
Considering Style

OBJECTIVES

★ Identify strategies for determining writing style.
★ Identify the formality of different genres.

BIG IDEA

As a writer, you must learn to adapt to your purpose or writing assignment while still being true to your own unique voice.

This lesson will discuss three factors that will help you determine style:

- Purpose and Audience
- Formality
- Complexity

Purpose and Audience

Some of the most common **purposes**, or goals, for writing include the following:

The ________________, or attitude you express in your writing, should be appropriate for both your ________________________ and your ________________________.

On Your Own

Imagine that you have been asked to write to a nursing class at the local college about vegetables. Identify the sentence below that would be a better addition to your paper:

☐ Vegetables are a yummy source of protein, which gives you super strong muscles.

☐ Vegetables can be a good source of protein, a nutrient that promotes muscle health.

Formality

Define the term **formality**.

Informal writing usually uses ________________-____________________ and ____________________-____________________ **pronouns** such as *I*, *me*, and *you*.

Formal writing is required for many ____________________________ or ____________________ genres. In these situations, you use more _______________________ sentence structures and ______________________________ terms and rarely use personal pronouns like *I* and *you*.

Complexity

Define the term **complexity**.

Lesson Wrap-up

Test Yourself

Next to each text category, write **F** for Formal or **I** for Informal. Check your work using the Answer Key in the back of the book.

1. _____ Research paper
2. _____ Business proposal
3. _____ Chapter on proper nursing procedures
4. _____ Blog post about the new restaurant in town
5. _____ Cover letter to a hiring manager
6. _____ Short story about your trip to Egypt
7. _____ Review of a book on Amazon
8. _____ Scholarly report on reading strategies

Key Terms

Beside each definition, write the corresponding term.

________________: The positive, negative, or neutral attitude that an author expresses about a topic

________________: The way a text conforms to certain standards

________________: The people who read your writing

Lesson 8R.2

Considering Visuals

OBJECTIVE

★ Understand how to effectively use visuals.

BIG IDEA

Learning how to analyze and effectively incorporate visuals into your writing can sharpen your critical thinking skills and strengthen your work.

In this lesson, you will learn how the purpose and context of images can affect an argument.

Purpose and Context

Purpose

To start ______________________ an image, you must first think about its ________________. Just like texts, images can ________________, ________________, ________________, or ________________.

All images make some kind of ________________________. Even ________________ photos and videos must ___________________ for your ______________________ and _______________________.

Keep in mind that the actual ______________________ of an image may be __________________________ from the stated purpose.

Context

Define the term **context**.

Lesson Wrap-up

Test Yourself

Next to each statement, write **T** for True or **F** for False. Check your work using the Answer Key in the back of the book.

1. ______ It's okay to upload pictures of your New York City vacation to your LinkedIn profile.
2. ______ You should avoid using any images in academic presentations because it's unprofessional.
3. ______ The purpose of an image can change based on its context.
4. ______ Only some images make an argument.
5. ______ Including appropriate, relevant images can strengthen your argument.

Key Terms

Write the number of each term next to its definition.

1. Purpose	______	How the contents of an image have been selected and arranged
2. Context	______	How and when information is presented
3. Argument	______	A reason why you should think a certain way
4. Entertaining Text	______	A text that shares a personal experience or belief
5. Informative Text	______	A text that gives an audience information about a topic
6. Persuasive Text	______	A text that explores a topic or event in a creative or humorous way
7. Reflective Text	______	A text that convinces its audience to adopt a belief or take an action

Chapter 9R
Integrated Review: Basics of Grammar & Mechanics

Lesson 9R.1
Nouns

OBJECTIVES

★ Define and identify nouns with different functions within a sentence.
★ Define and identify types of nouns.

BIG IDEA

A noun is a part of speech that represents a person, place, thing, event, or idea; the English language is full of them, and without them, you wouldn't be able to talk about much.

Fill in the table with your own examples of nouns:

People	Places	Things	Events	Ideas

In this lesson, you will learn about the following:

- Common and Proper Nouns
- Singular and Plural Nouns
- Count and Non-Count Nouns
- Compound Nouns
- The Functions of Nouns

On Your Own

Read the following paragraph and identify a noun.

> Africa has seen many great leaders, but none have been as influential as Nelson Mandela. He fought against Apartheid, a political revolution that violently divided people by race. Mandela is best known for serving time in jail as a result of political protesting. The country mourned his death when he passed in South Africa in 2013.

Common and Proper Nouns

Proper nouns are the names of ____________________ people, places, things, events, or ideas.

Common nouns name _____________-______________________ people, places, things, events, or ideas.

On Your Own

In the table below, use the list of common nouns to create your own list of proper nouns. For example, for the common noun *city*, you could write *Seattle* or *Budapest*. Write your ideas in the spaces provided.

Common Nouns	Proper Nouns
author	
activist	
friend	
state	
school	
website	
dog	

Singular and Plural Nouns

Singular nouns refer to ________________________ person, place, thing, event, or idea. **Plural nouns** refer to ________________________ people, places, things, events, or ideas.

Complete the following table with your own examples of common, proper, singular, and plural nouns.

Common Nouns	Proper Nouns	Singular Nouns	Plural Nouns
college	Baylor University	Canada	countries

Count and Non-Count Nouns

Count nouns are nouns that can be ______________________________. **Non-count nouns** are nouns that can't be __________________________.

Which of the following nouns are count and which are non-count? Write *count* or *non-count* in the table below.

three mails	
three Targets	
three electricities	
three concerts	

Compound Nouns

Nouns that are made up of more than one word are known as ____________________________ ______________________. Some compound nouns are joined together into one word while some are separated by ______________________ or ______________________.

The Functions of Nouns

Subject

Nouns can act as the **subject** of a sentence. The subject of a sentence is simply ______________________ or __________________ the sentence is about.

Grammar Practice 1

Identify the noun acting as the subject in each sentence. Check your work using the Answer Key in the back of the book.

One lucky student will win this awesome bike.

My sister always gets good grades.

Add a subject to each sentence. Check your work using the Answer Key in the back of the book.

My ________________ called me last night at 11:00.

Several _______________ crossed the road in front of me.

Direct Object

A **direct object** receives the ____________________ of a **verb**.

Grammar Practice 2

Identify the direct object in each sentence. Check your work using the Answer Key in the back of the book.

The chicken ate a bug.

The farmer planted his crops.

Object of a Preposition

The **object of a preposition** is the noun that completes a ______________ ____________________.

(Prepositions show location, direction, or time.)

Grammar Practice 3

Identify each noun acting as an object of a preposition in each of the following sentences. Check your work using the Answer Key in the back of the book.

My nephew hid under the table and wanted me to find him.

We went down the river and swam in the deep water.

Adjectives

Nouns can act like **adjectives**, describing other _________________ or ____________________.

Grammar Practice 4

Identify the noun acting as an adjective. Check your work using the Answer Key in the back of the book.

The mail truck has arrived!

I can't believe the cat knocked over his own food bowl.

Lesson Wrap-up

Key Terms

Define the following Key Terms from this lesson.

Adjective:

Common Noun:

Compound Noun:

Count Noun:

Direct Object:

Hyphen:

Non-count Noun:

Noun:

Object of a Preposition:

Plural Noun:

Prepositional Phrase:

Proper Noun:

Singular Noun:

Subject:

Verb:

Lesson 9R.2

Pronouns

OBJECTIVES

★ Define different types of pronouns and related terms.
★ Identify indefinite pronouns.
★ Identify relative pronouns.
★ Locate and classify personal pronouns by person, gender, number, and case.
★ Understand the relationship between pronouns and antecedents.

BIG IDEA

A pronoun is a word that takes the place of a noun in a sentence and can represent people, places, things, events, or ideas. Understanding and utilizing pronouns helps you avoid repeating the same word over and over again in the same piece of writing.

In this lesson, you will learn about the following:

- Personal Pronouns
- Indefinite Pronouns
- Relative Pronouns
- Pronouns in Sentences

Personal Pronouns

Personal pronouns rename specific ________________, ________________, ____________________,

or ____________________.

Pronouns can be categorized in four ways:

1.
2.
3.
4.

Grammar Practice 1

Identify all the personal pronouns in the sentences below. Check your work using the Answer Key in the back of the book.

After soccer practice, I made sure to drink plenty of water.

You aren't really a true Jacksonville Jaguars fan unless you get upset when they lose!

Number: Plural and Singular Pronouns

When pronouns are divided by **number**, they are separated into two groups: ________________ and ________________. Singular pronouns refer to one person or object while plural pronouns refer to multiple people or objects.

Grammar Practice 2

Locate the pronouns provided for each of the following sentences. Check your work using the Answer Key in the back of the book.

Identify all the singular pronouns in the following sentence:

> I was in the middle of cooking steak and potatoes when my friend invited me out to dinner.

Identify all the plural pronouns in the following sentence:

> The group decided that we would go for a walk after dinner; our stomachs needed help digesting.

An __________________ is the word a pronoun renames in a sentence.

Person: First-, Second-, and Third-Person Pronouns

Use **first-person pronouns** when talking about ____________________.

Example:

> I walked alone.

Use **second-person** pronouns when talking directly to ______________________.

Example:

> Will you run to the grocery store?

Use **third-person** pronouns for ____________________ __________________.

Example:

> She is coming to the baseball game.

Grammar Practice 3

In the sentences below, identify the pronouns based on their category. Check your work using the Answer Key in the back of the book.

First Person

> What is up with me these days? I have been so scatterbrained lately that my friends are concerned.

Second Person

> What are your plans later on today? If you would like, we can go shopping.

Third Person

> She has been doing such a great job at work that they have promoted her to manager.

Gender: Male, Female, and Neutral Pronoun Genders

There are three genders in the English language: _______________, _______________, and ___________________.

Male-gender pronouns:

Female-gender pronouns:

Neutral-gender pronouns:

Grammar Practice 4

Complete the following sentences with the correct pronoun. The gender of the pronoun should match the traditional gender of the noun it renames. Check your work using the Answer Key in the back of the book.

Franco washed ______________ car.

Cindy said that the coat belongs to ______________ .

Our neighbors invited us to ______________ house for dinner.

Case: Subjective-, Objective-, and Possessive-Case Pronouns

Subjective-case pronouns are used as the ________________ of a ____________________.

Objective-case pronouns are used as objects of ________________, ________________

______________________, or ____________________ ______________________.

Possessive-case pronouns are used as ______________ ________________.

Grammar Practice 5

Complete each sentence with an appropriate pronoun based on the directive provided. Check your work using the Answer Key in the back of the book.

That is ___________ car.	(Possessive-case pronoun)
Will you come to the movies with mom and ___________?	(Objective-case pronoun)
___________ went to the store for me.	(Subjective-case pronoun)
___________ car is parked in ___________ parking space.	(Two possessive-case pronouns)
I threw the ball to ___________.	(Objective-case pronoun)

Indefinite Pronouns

Indefinite pronouns are pronouns that don't ______________ a ______________ ______________.

Make a list of eight indefinite pronouns:

1. 3. 5. 7.

2. 4 6. 8.

Relative Pronouns

List the seven relative pronouns:

1.
2.
3.
4.
5.
6.
7.

Relative pronouns are used to ______________________ **dependent clauses**, groups of words with a subject and a **verb** that do not __________________ a complete ________________.

Grammar Practice 6

Identify the relative pronoun in the sentence below. Check your work using the Answer Key in the back of the book.

> At the time of the election, many people were unhappy with both candidates, which led to political unrest.

Case

___________ and ______________ are the only relative pronouns that have a specific case. *Who* is a _____________________-case pronoun and *whom* is an _____________________-case pronoun.

Pronouns in Sentences

Pronoun Reference

When used in sentences, pronouns *refer* to __________________ or other _______________________.

Here's an example of *unclear* reference:

> They have been working on them for a while, and they should be done by the end of the day.

If the scenario is talking about employees getting their expense reports updated, how can you edit the sentence for clearer pronoun reference? Rewrite the sentence below.

Pronouns and Antecedents

Pronouns must agree with their antecedents in __________________, __________________, and ______________________.

Lesson Wrap-up

Test Yourself

Identify the pronouns in the following sentences and indicate their cases (subjective, objective, or possessive). If applicable, indicate whether each pronoun is indefinite or relative. Check your work using the Answer Key in the back of the book.

1. Whenever I feel sad, I listen to my favorite music.
2. They are coming with you and me to the party.
3. Did you go to their house for Thanksgiving dinner?
4. My sister made up a game that was too difficult to play.
5. Please tell whomever is on the phone to call back later.

Key Terms

Define the following Key Terms from this lesson.

Antecedent:

Direct Object:

First-person Pronoun:

Gender:

Indefinite Pronoun:

Number:

Person:

Personal Pronoun:

Pronoun:

Pronoun-Antecedent Agreement:

Pronoun Case:

Pronoun Reference:

Relative Pronoun:

Second-person Pronoun:

Third-person Pronoun:

Lesson 9R.3

Basic Verb Types and Tenses

OBJECTIVES

- ★ Identify and understand the characteristics and purposes of verbs.
- ★ Recognize types of verbs.
- ★ Recognize verbs in the simple tenses.

BIG IDEA

Verbs are words that represent actions, relationships, or states of being. Verbs are an essential part of every sentence, and it would be impossible to communicate without verbs, which hold everything together.

In this lesson, you w ll learn about the basic verb types and tenses:

- Verb Types
- Simple Verb Tenses

Verb Types

Action Verbs

Action verbs show ________________ or ________________ ______________________.

Example:

On her birthday, Stella opened her presents with joy.

Direct & Indirect Objects

Sometimes, an action verb is followed by a __________ ____________, a word that receives the action of the verb.

Example:

Gina accidentally *baked* the pie for too long.

An **indirect object** is a word that receives the direct object.

Example:

After the semester, the professor *gave* Alonzo a **letter** of recommendation.

Linking Verbs

Linking verbs link the ___________________ to a ____________________.

List eight examples of linking verbs:

1.	3.	5.	7.
2.	4.	6.	8.

Linking Verbs vs Action Verbs

There are some linking verbs that can also function as action verbs. To determine which way they are used, you should ____________ the _______________________ of the sentence.

Examples:

Darla tasted the food.

The food tasted awful.

List ten words that commonly function as both action and linking verbs:

1.
2.
3.
4.
5.
6.
7.
8.
9.
10.

Helping Verbs

The **main verb** is the verb that expresses the primary action or state of being in a complete sentence. **Helping verbs** are added to a main verb to create a new verb that grammatically fits the sentence.

Example:

Jenny has completed all training in preparation for next month's marathon.

In this sentence, the helping verb *has* is helping the main verb *completed*.

Helping Verbs vs Linking Verbs

Some helping verbs can also be used as both helping verbs and linking verbs.

Examples:

The lifeguard was watching the beachgoers during his entire shift.

In this sentence, the verb ____________ is helping the main verb ________________.

The lifeguard was happy to be off work after a long day in the sun.

In this sentence, the verb *was* is linking the subject, _________________, to its description: ________________.

On Your Own

Read the following paragraph and identify a helping verb.

The marketing company was using both online and offline strategies to reach its audience. Usually, this audience is unified by a common interest or goal, not by the corporation itself. Because the company was planning events, promotions, and contests, the audience was more trusting. This type of marketing is often wildly successful; in many instances, the company even encourages new product development for the corporation.

Grammar Practice 1

Fill in the blanks with a helping verb to make the sentence grammatically correct. Check your work using the Answer Key in the back of the book.

She ___________ take piano lessons next year.

I __________ going to the mall right now.

Simple Verb Tenses

Verbs also have ______________ that let the reader know *when* an action is taking place. The most common tenses are _____________ tense, _____________ tense, and _____________ tense.

Past Tense

Past tense is used to report an event or reflect on a ____________________ experience.

Examples:

Past-tense action verb:

Frederick borrowed his brother's iPhone charger yesterday.

Past-tense helping verb:

Jamie was uninterested in eating meat after watching a documentary on animal cruelty.

Past-tense action verb (with helping verb):

Prasad was walking to work.

Regular vs Irregular Verbs

With regular verbs, the past tense is formed by _________________ *–ed* (or *–d* if the verb ends in *e*) to the end of the base form of the verb.

Example:

I listened to my professor talk about the Civil War last week.

Irregular verbs have unique patterns and can completely change form in the past tense.

Example:

Ryan drove his mother to the airport yesterday.

On Your Own

Identify a past-tense verb in the following paragraph.

Cheryl was going to go to the store after work, but the storm stopped her. Before she knew it, all kinds of debris went flying past her office window. Hail rained down on her car parked just ten feet away, and garbage from the nearby trashcan stuck to its window. She was glad Jennifer had stopped her to discuss the meeting next week, or she would have been caught in that storm!

Present Tense

When writing about events or actions that are happening __________, you should use the **present tense**.

Examples:

Present-tense action verb:

Alyssa watches her German Shepherd play at the dog park.

Present-tense linking verb:

The new restaurant is affordable.

Present-tense action verb (with helping verb):

Westley is thinking he might get into law school next year.

On Your Own

Read the following paragraph and identify a present-tense verb.

The air is frigid, and my hands are numb. I think even the snow foxes are huddled together for warmth. This morning was warmer, but now I am sure the temperature has dropped at least fifteen degrees. My dog stays close to my side and whines to find warmth and shelter. The sun is starting to break through the clouds, but somehow it seems colder than it was before.

Future Tense

The **future tense** is used to describe events or actions that have not yet __________ __________ or to describe plans or instructions.

Example:

This living room will look incredible once it is furnished.

On Your Own

Read the following paragraph and identify a future-tense verb.

> One day, she will look up and see just how far she has come. She will realize that everything has been building to this moment. She will gaze out over the crowd of people cheering her name, and the weight of ten years' hard work will disappear.

Lesson Wrap-up

Key Terms

Define the following Key Terms from this lesson.

Action Verb:

Adverb:

Direct Object:

Future Tense:

Helping Verb:

Indirect Object:

Linking Verb:

Main Verb:

Past Tense:

Prepositional Phrase:

Present Tense:

Subject:

Verb:

Lesson 9R.4

Verb Forms and Functions

OBJECTIVE

★ Identify and understand verbals.

BIG IDEA

The base form of a verb can be altered or expanded to create verbals. Being familiar with the three primary types of verbals will help you write clear and effective sentences.

Define the term **verbal**.

In this lesson, you will learn about three types of verbals:

- Infinitives
- Gerunds
- Participles

Infinitives

The word *to* + the base form creates the **infinitive**: a verbal that can function as a noun, adjective, or adverb in a sentence. Infinitives are often used to reference expectations or intentions.

Example:

Remind me not <u>to forget</u> my water bottle when we leave the gym.

Gerunds

Gerunds function as nouns and are formed by adding *–ing* to the base form of a verb.

Example:

<u>Diving</u> is Jenna's favorite sport to watch.

Participles

Past Participles

The **past participle** is a verb form used to show completed mental or physical action. Past participles can function as adjectives that express an action or a state of being.

Example:

Dale the dog was <u>excited</u> to receive a treat after a long day of being a good boy.

Regular vs Irregular Past Participles

A regular past participle is formed by adding *–ed* (or just *–d* if the verb ends in *e*) to the end of a regular verb. Irregular verbs change completely to form their past participles.

What is the regular past participle of the verb *walk*?

What is the irregular past participle of the verb *catch*?

Present Participles

A **present participle** is formed by adding *-ing* to the ________________ of the base form of a verb.

Present participles can function in two ways:

1.
2.

On Your Own

Practice forming present participles with the following verbs by filling in the blank column. The first example has been completed for you.

Base Form	Present Participle
cook	cooking
smell	
write	
break	
sing	

Lesson Wrap-up

Key Terms

Define the following Key Terms from this lesson.

Base Form:

Gerund:

Infinitive:

Past Participle:

Present Participle:

Verbal:

Lesson 9R.5

Perfect and Progressive Tenses

OBJECTIVES

★ Recognize verbs in the perfect and progressive tenses.

BIG IDEA

The perfect and progressive tenses help you speak more specifically about when actions or events occur.

In this lesson, you will learn about the following verb tenses:

- Perfect Tense
- Progressive Tense

Perfect Tense

Past-Perfect Tense

The **past-perfect tense** is used to describe a ________________ action that was ________________ before another past-completed action.

Example:

Amanda had driven forty miles before she realized she forgot her purse at home.

Present-Perfect Tense

The **present-perfect tense** is used to describe:

It is formed by combining the correct form of the helping verb *have* with a verb's ________ __________.

Future-Perfect Tense

The **future-perfect tense** ____________ an ________________ that will be completed before another ________________ event. It is formed by ________________ *will have* with the past participle of a ________________.

On Your Own

Read the following sentences and choose the box to indicate whether the sentence is in past-perfect tense, present-perfect tense, or future-perfect tense.

Shelia has worked hard over the past fifteen years.

☐ Past-perfect tense
☐ Present-perfect tense
☐ Future-perfect tense

Alyssa had tried to talk to Marion several times before.

- ☐ Past-perfect tense
- ☐ Present-perfect tense
- ☐ Future-perfect tense

I will have watched ten movies by the end of spring break.

- ☐ Past-perfect tense
- ☐ Present-perfect tense
- ☐ Future-perfect tense

Progressive Tense

Past-Progressive Tense

The **past-progressive tense** is used to describe actions that meet one of three criteria:

1. actions that were already in progress in the past

2.

3.

The past-progressive tense is formed by using the past-tense form of the verb *be* with the present participle of a verb.

Example:

I was cleaning the kitchen when my friend came over for coffee.

Present-Progressive Tenses

The **present-progressive tense** is used to:

1. describe a recurring action

2.

3.

It is formed using the present-tense form of the verb *be* and the present participle of a verb.

Example:

The chefs are preparing the buffet dinner for tonight's event.

Future-Progressive Tense

The **future-progressive tense** is used to describe ____________________ that ______________ happen in the ___________. It is formed by combining the verb phrase *will be* and the ____________________ ________________ of the verb.

Example:

Beyoncé <u>will</u> <u>be</u> <u>headlining</u> after the opening artist performs.

On Your Own

Identify the present participle in each of the following sentences. Then, check the box next to the correct tense.

The storm is going to be intense this weekend.

- ☐ Past-progressive tense
- ☐ Present-progressive tense
- ☐ Future-progressive tense

She was eating her dinner when the call came in.

- ☐ Past-progressive tense
- ☐ Present-progressive tense
- ☐ Future-progressive tense

I will be studying that chapter tonight.

- ☐ Past-progressive tense
- ☐ Present-progressive tense
- ☐ Future-progressive tense

Lesson Wrap-up

Key Terms

Define the following Key Terms from this lesson.

Future-perfect Tense:

Future-progressive Tense:

Past-perfect Tense:

Past-progressive Tense:

Present-perfect Tense:

Present-progressive Tense:

Lesson 9R.6

Adjectives and Adverbs

OBJECTIVE

★ Understand the characteristics and purposes of adjectives and adverbs.

BIG IDEA

Adjectives and adverbs are especially helpful for giving descriptions and adding detail.

In this lesson, you will learn about:

- Forms and Functions of Adjectives
- Forms and Functions of Adverbs

Forms and Functions of Adjectives

Adjectives are used to describe ____________________ or ____________________. Adjectives answer these five questions:

1.

2.

3.

4.

5.

When you are comparing two items, follow this rule:

When you are comparing more than two items, follow this rule:

On Your Own

Identify the adjectives (excluding articles) in the following sentences:

The play expertly portrays the complicated and intricate web of relationships.

We have just received your letter concerning the upcoming publication of your new novel.

This article thoroughly examines the role of technology in strong reading and writing skills.

Forms and Functions of Adverbs

Adverbs describe ____________________, ______________________, and ____________________

______________________.

Adverbs answer these five questions:

1.
2.
3.
4.
5.

Grammar Practice 1

Identify the adverb in each sentence. Check your work using the Answer Key in the back of the book.

The elderly man limped painfully after his surgery.

I shop at that store often.

Mom hid her jewelry carefully.

Just like adjectives, adverbs can be used to make ______________________________. To compare two actions with an adverb, follow this rule:

On Your Own

Read the following sentences and identify adverbs.

The play expertly portrays the complicated and intricate web of family relationships.

We have just received your letter concerning the upcoming publication of your new novel.

This article thoroughly examines the role of technology in strong reading and writing skills.

Lesson Wrap-up

Key Terms

Define the following Key Terms from this lesson.

Adjective:

Adverb:

Article:

Linking Verb:

Noun:

Pronoun:

Lesson 9R.7

Prepositions

OBJECTIVE

- ★ Understand the characteristics and purposes of prepositions.

BIG IDEA

Prepositions are words that show relationships among people, places, things, events, and ideas.

In this lesson, you will learn about:

- Purposes and Functions of Common Prepositions
- Purposes and Functions of Prepositional Phrases

Purposes and Functions of Common Prepositions

Prepositions are words that show ______________________ among the words in a

____________________.

Grammar Practice 1

Fill in the blanks with a preposition that shows location.

We took a ride ______ the boat.
We went ______ the opera house.
Sarah met her friends ______ the park.

Prepositions can also show relationships such as time. Fill in the blanks with a preposition that shows time:

Please have dinner ready ________ 6:00.
I drank a tall glass of water ________ the game.
I haven't seen her ________ eighth grade.

List 25 of the most common prepositions:

Purposes and Functions of Prepositional Phrases

Prepositional phrases are _______________ of _______________ words that begin with a preposition and end with a noun or pronoun. The noun or pronoun that ends a prepositional phrase is known as the _______________ **of the preposition.**

In a sentence, prepositional phrases usually function as either _______________ or _______________.

Prepositional Phrases as Adjectives

If a prepositional phrase is an adjective, it _______________ a _______________ or _______________ and answers several questions.

Write the five questions that a prepositional phrase being used as an adjective will answer:

1.

2.

3.

4.

5.

On Your Own

Read the sentence and select the prepositional phrase acting as an adjective.

I went to the office so that I could borrow that important file of Jimmy's.

Prepositional Phrases as Adverbs

A prepositional phrase that is acting as an adverb will describe a _______________ and answer one of the following questions:

1.

2.

3.

4.

5.

Lesson Wrap-up

Key Terms

Match each Key Term with its definition.

A. Object of the Preposition
B. Adverb
C. Pronoun
D. Prepositional Phrase
E. Noun
F. Adjective
G. Preposition
H. Subject
I. Verb

_____ A word that describes a noun or pronoun

_____ A word that describes or modifies a verb, an adjective, or another adverb

_____ A word that names a person, place, thing, event, or idea

_____ A word that completes the meaning of a preposition in a prepositional phrase

_____ A word used as a substitute for a noun

_____ A word that shows a relationship among people, places, things, events, and ideas

_____ The preposition with its object and any modifiers

_____ Who or what the sentence is about

_____ A word that represents an action, relationship, or state of being

Lesson 9R.8

Clauses and Conjunctions

OBJECTIVES

- ★ Identify clauses and understand ways to combine them.
- ★ Identify different types of conjunctions.
- ★ Recognize and understand conjunctive adverbs.
- ★ Understand the characteristics and purposes of conjunctions.

BIG IDEA

Your writing will become smoother and easier to read when you utilize clauses and conjunctions to join ideas together in a logical way.

In this lesson, you will learn about forming simple, compound, complex, and compound-complex sentences with the following components:

- Related Words and Clause
- Coordinating Conjunctions
- Subordinating Conjunctions
- Correlative Conjunctions
- Conjunctive Adverbs

Related Words and Clauses

A **simple sentence** includes the following three elements:

1.

2.

3.

Clauses

An **independent clause** can stand alone as a ____________________ ______________________.

A **dependent clause** is a word group containing a ________________ and a ________________ that cannot stand alone as a complete sentence.

Sentence Patterns

List the combinations in each sentence pattern:

Simple Sentence	
Compound Sentence	

Complex Sentence	
Compound-Complex Sentence	

Define the term **conjunction.**

List the three main types of conjunctions:

1.
2.
3.

Coordinating Conjunctions

Coordinating conjunctions join similar ______________ or ______________ of words together.

List the seven coordinating conjunctions:

F	
A	
N	
B	
O	
Y	
S	

Words

The first function of coordinating conjunctions is joining two or more words, such as ________________, ________________, ________________, or ________________.

Phrases

Coordinating conjunctions can also be used to join whole phrases, such as ________________________ ______________________.

Clauses

Finally, coordinating conjunctions help form compound sentences. A **compound sentence** is a sentence pattern in which two __________ __________ are connected by a **comma** and a coordinating conjunction.

On Your Own

Read the following passage and identify a coordinating conjunction.

> Every day, Mary and Susan leave their houses, and they walk to school together. The neighbor's dog usually follows them, but today he was nowhere to be found. Susan wondered if the dog was sick, so before they went to school, they stopped at the neighbor's house. The dog poked his nose out when his owner answered the door, and the girls smiled as they went off in the direction of the school.

Subordinating Conjunctions

Subordinating conjunctions introduce dependent clauses. A dependent clause can join an independent clause to form a __________ __________.

List ten common subordinating conjunctions:

1.	3.	5.	7.	9.
2.	4.	6.	8.	10.

On Your Own

Read the following passage and identify a subordinating conjunction.

> While Sara was doing the dishes, she heard a knock on her door. She went to answer it even though she wasn't expecting anyone. When she opened the door, a cat peered up at her. It looked hungry and dirty and had, apparently, learned to knock. It wasn't until Sara stepped outside that she saw the pot that the cat had been rubbing against to knock on the door.

Correlative Conjunctions

Correlative conjunctions are similar to __________ __________; they both connect two or more word groups of __________ __________.

Correlative conjunctions always appear in __________.

Write the five correlative conjunction pairs:

1.
2.
3.
4.
5.

Correlative Conjunctions and Subject-Verb Agreement

When using *both/and* to join subjects in a sentence, make sure you use a ______________ verb. When using *either/or* to join subjects in a sentence, make sure the verb agrees with the subject ____________________ to it.

On Your Own

Read the following passage and identify the sentence that uses correlative conjunctions.

> Seattle, Washington is appealing because it provides access to mountains and beaches. Downtown Seattle, including the famous Pike Place Market, is right on the water. Originally, Pike Place was a fish market. Today, it's a maze of both indoor and outdoor shops. These shops sell everything from spices to flower bouquets, and of course fresh seafood. There are also delicious restaurants and cafés in this area. However, if you visit Pike Place, be prepared for a crowd.

Conjunctive Adverbs

Like conjunctions, **conjunctive adverbs** are used to show ______________________, ______________________, ____________________________, and other relationships between clauses.

Add a conjunctive adverb to the following sentence:

> Winnie didn't finish her chores; she wasn't allowed to go to the beach.

Lesson Wrap-up

Test Yourself

Follow the instructions to select the appropriate type of conjunctions. Check your work using the Answer Key in the back of the book.

Identify the coordinating conjunctions in the following sentences:

1. I wanted ham, but we didn't have any, so we went out to eat instead.
2. We had ice cream and cake at the birthday party, but not everybody wanted some.
3. He and I will be getting married next year, so we have started planning now.

Identify the subordinating conjunctions in the following sentences:

1. Ever since I started exercising, I have felt a lot better.
2. After snacking all day, I couldn't eat dinner.
3. In spite of our differences, we have remained friends since first grade.

Identify the coordinating conjunctions that are joining two prepositional phrases:

1. We went over the mountain and through the woods.
2. I went through the yard and into the house.

Key Terms

Define the following Key Terms from this lesson.

Comma:

Complex Sentence:

Compound Sentence:

Compound-complex Sentence:

Conjunction:

Conjunctive Adverb:

Coordinating Conjunction:

Correlative Conjunctions:

Dependent Clause:

FANBOYS:

Independent Clause:

Prepositional Phrase:

Semicolon:

Sentence Patterns:

Subordinating Conjunction:

Lesson 9R.9

Using Capitalization and Italics

OBJECTIVE

★ Understand the purposes of capitalization and italics.

BIG IDEA

Capitalization and italics are both used in writing as visual markers to set apart words and phrases. They should be used sparingly, but these helpful tools remind your audience to pay close attention, as important or unique information is to come.

In this lesson, you will learn about how to correctly use the following:

- Capitalization
- Italics

Capitalization

Write the six most common uses of capital letters:

1.
2.
3.
4.
5.
6.

Sentences

You should always capitalize the ________ ____________ in any sentence. If you are using an **inline quotation**, you do not need to use a capital letter at the beginning.

Proper Nouns

Capitalize **proper nouns**, which are words that name a specific ______________, ______________, ______________, ______________, or ______________.

On Your Own

Read the examples and fill in your own.

Proper Nouns	Examples	Add Your Own
People	George Takei Una Thompson	
Places	Mount Kilimanjaro Brazil	

Organizations	Kraft Foods Harvard University	
Days, months, and holidays	Monday Yom Kippur	
Historical events and eras	the Civil War Prohibition	
Ethnicities and nationalities	Asian American Cherokee	
Religions and deities	Hinduism Christianity	
Titles	"The Raven" *The Scarlet Letter*	

Other Uses

Capitalization should also be used in the following situations:

Italics

Italics are ______________________ letters that are used to __________________ __________________ certain words or phrases.

Italics are generally used for:

What are the seven types of longer works that should be italicized?

1.
2.
3.
4.
5.
6.
7.

States	MN SC PA	

Numbers

In math class, you will almost always use ______________________, like 9 or 67, when you are talking about numbers. In an English paper, however, you would _______________ those same numbers out as ___________ or ___________-____________. You should also use words to ________________ a larger number (like $3.5 million.)

When should you use actual numbers, or numerals?

On Your Own

Read the paragraph below and identify a sentence that uses numbers correctly.

> The 4 company headquarters owned by Liberty, Inc. house a wide variety of careers and professional opportunities. There are a total of twenty thousand employees and ninety-five management positions. Each headquarters contains roughly twenty-five departments. Their customer base is large as well; they serve a population of 1,500,000 clients.

Lesson Wrap-up

Key Terms

Define the following Key Terms from this lesson.

Abbreviation:

Hyphen:

Parentheses:

Phrase:

Lesson 9R.11

Using Basic Spelling Rules

OBJECTIVES

- ★ Understand basic spelling rules.
- ★ Understand how to spell *ie* and *ei* words.
- ★ Understand how to spell plural words.
- ★ Understand how to spell words with final consonants.
- ★ Understand how to spell words with suffixes.

BIG IDEA

Correct spelling is crucial in academic and professional settings if you want to make a good impression.

In this lesson, you will learn useful spelling rules for the following situations:

- *IE* and *EI* Words
- Plural Words
- Suffixes

IE and *EI* Words

Fill in the table with examples from the lesson.

ie	After *c*	Words like *weigh*

Exceptions

Use *ie* after the *-sh* sound:

Use *ei* if the vowels are pronounced like the *i* in *bit*:

Use *ei* in abnormal words:

On Your Own

Choose the correctly spelled word. Rewrite the others with the correct spelling of each.

- ☐ Reciept
- ☐ Ancient
- ☐ Foriegn
- ☐ Releif

Plural Words

When you are making a word plural, pay attention to the ____________________ of the word. Add *-es* to any words that end with the following letters:

If a word ends in any other letter, follow this rule:

On Your Own

Read the table of plural words below and fill in the blanks with your own examples.

-es	-s
mess → messes	assignment → assignments
brush → brushes	caterpillar → caterpillars
church → churches	nap → naps
box → boxes	phone → phones
maze → mazes	scissor → scissors

Words that end in _________ are a special exception to this rule. Some are spelled with *-es* and some are spelled with *-s*.

On Your Own

Read the table of plural words that end in *-o* below and fill in the blanks with your own examples.

-es	-s
echo → echoes	combo → combos
hero → brushes	logo → logos

potato → potatoes	taco → tacos
tomato → tomatoes	typo → typos
veto → vetoes	video → videos

Finally, remember that some words ____________________ their spelling or ______________ ___________ __________________ when they are made plural.

On Your Own

Read the tables below and fill in the blanks with your own examples.

Singular	**Plural (Spelling Changes)**
child	children
goose	geese
mouse	mice
person	people

Singular	**Plural (Spelling Stays the Same)**
deer	deer
fish	fish
sheep	sheep
species	species

Suffixes

Suffixes are _______________ _______________ added to the end of a root in order to change the meaning.

Silent *-e*

Many words in the English language end in a silent *-e*. If you are adding a suffix that starts with a vowel to the word, drop the silent *-e*.

On Your Own

Read the table below and fill in the blanks with your own examples.

Original	Vowel Suffix	Loses *e*
dance	-ing	dancing
narrate	-ion	narration
excite	-able	excitable
file	-ed	filed

On Your Own

Read the table below and fill in the blanks with your own examples.

Original	Vowel Suffix	Loses *e*
advertise	-ment	advertisement
hope	-ful	hopeful
safe	-ty	safety
state	-ly	stately

Final -*y*

When the letter before the final -*y* is a vowel, follow this rule:

On Your Own

Read the table below and fill in the blanks with your own examples for different final *-y* scenarios.

Original	Suffix	New
employ	-er	employer
enjoy	-ing	enjoying
try	-s	tries
mystery	-ous	mysterious

Final Consonants

When adding a ____________________ to a one-syllable word that ends with a *consonant-vowel-consonant* pattern, you must ________________________ the final consonant.

Examples:

jam → jamming
snap → snapped

On Your Own

Read the table below and fill in the blanks with your own examples for different final consonant scenarios.

Original	New
sit	sitting
knot	knotting
refer	referred
submit	submitting

Lesson Wrap-up

Key Terms

Define the following Key Terms from this lesson.

Stress:

Suffix:

Syllable:

Lesson 9R.12

Spelling Commonly Confused Words

OBJECTIVES

- ★ Learn how to spell commonly confused words.
- ★ Learn how to spell commonly misspelled words.

BIG IDEA

Learning the definitions and functions of confusing words will help you utilize them properly.

This lesson will teach you the differences between two types of commonly misused words:

- Similar Sound Words
- Similar Meaning Words

Similar Sound Words

Write a brief definition or description that will help you remember the meaning of each word:

Accept:	Except:
Its:	It's:
Choose:	Chose:
Past:	Passed:
There:	Their:

Affect:	Effect:
Loose:	Lose:
Its:	It's:
Than:	Then:
They're:	Too:

To:	Two:	Whether:	Weather:
Whose:	Who's:	Your:	You're:

Similar Meaning Words

Words with similar meanings are called ____________________. Words with opposite meanings are called ____________________.

Write a brief definition or description that will help you remember the meaning of each word:

Between:	Among:	Fewer:	Less:
Come:	Go:	Lie:	Lay:

On Your Own

Read the following sentences and determine which word fits in the blank.

1. Let's meet at (your / you're) place to work on the group project.
2. The squirrel stopped in (its / it's) tracks in the middle of the sidewalk.
3. When it's hot outside, wear (loose / lose) clothing and drink water regularly.
4. My friend (lends / borrows) pens from me in class and never gives them back.
5. The article claimed that (less / fewer) music has been illegally downloaded this year.
6. Every day at three o'clock, my grandfather (lays / lies) down to take a nap.

Lesson Wrap-up

Test Yourself

Choose the correct word to complete each sentence. Check your work using the Answer Key in the back of the book.

1. Do not (accept / except) friend requests from people you don't know.
2. How did that medicine (effect / affect) you?
3. The mouse hid (its / it's) cheese in the hole in the wall.
4. I love studying about our nation's (past / passed).
5. I could not get the knots in my shoelaces (loose / lose).
6. Which kind of cheesecake did you (choose / chose)?
7. I'd rather go swimming (than / then) do my chores.
8. Her house is over (there / their).
9. (Their / They're) moving into a new home.
10. Would you like to come with us, (to / too)?
11. We are going (to / too) the movies.
12. Oh, the (whether / weather) outside is frightful.
13. Have you found out (whose / who's) keys those are?
14. (Your / You're) going to get into trouble if you don't stop it.
15. Just (between / among) you and me, I hated that cake.
16. Will you please (come / go) over here and help me?
17. There are (fewer / less) students in my class this semester.
18. I have a headache, so I will (lay / lie) here for a while.

Key Terms

Define the following Key Terms from this lesson.

Antonym:

Homonym:

Synonym:

Chapter 10R
Integrated Review: Grammatical Sentences

Lesson 10R.1
Identifying the Characteristics of Sentences

OBJECTIVES

- ★ Identify the characteristics of complete sentences.
- ★ Understand the characteristics and purposes of dependent and independent clauses.
- ★ Understand the characteristics and purposes of parts of speech.
- ★ Recognize dependent and independent clauses in sentences.

BIG IDEA

The ability to group words together in complete sentences is essential in meaningful communication.

In this lesson, you will learn about the following sentence components:

- Capitalization and Punctuation
- Parts of Speech
- Subject and Predicate
- Complete Thought

Capitalization and Punctuation

Complete sentences always start with a ____________ ____________ and end with a ____________, a ____________ ____________, or an ____________ ____________.

Period:

Question mark:

Direct question:

Indirect question:

Exclamation point:

Types of Statements

Declarative sentences make ____________________ statements that inform the reader and end in a ____________________.

Interrogative sentences ask ____________________ that end with a ____________________ mark.

Imperative sentences make commands and often address an ______________________ (*you*) ________________. They can end with a ____________________ or an _____________________ ________________.

Exclamatory sentences communicate _______________________ with _________________ or _____________________, and they end with __________________________ points.

Parts of Speech

Nouns and Pronouns

A **noun** is a word that represents a ______________________, ______________________, ____________________, __________________, or __________________.

A **pronoun** is a part of speech used to replace a ___________________ or another __________________.

Classifying Nouns

Common Noun:

Proper Noun:

Count Noun:

Non-count Noun:

Classifying Pronouns

Personal Pronoun:

Relative Pronoun:

Indefinite Pronoun:

On Your Own

Identify the nouns in the following sentences:

> Ryan and Camilla went to the movie theater to see Star Wars after eating sushi. Next weekend, they'll go to Savannah for their anniversary.

Identify the pronouns in the following sentences:

> Can you come to the concert tomorrow? Jenna is sick; since she can't make it, my extra ticket is now available!

Verbs

Verb Types

Action verbs show ______________________ or __________________ action.

Linking verbs link the ___________________ to a _____________________________.

Helping verbs change the _______________ of the main ______________ to grammatically ___________ into a sentence.

Verb Tenses

The **past tense** is used to __________________ or __________________ on events that took place in the ______________.

The **present tense** is used to describe __________________ or ________________ that are happening ________ or to reference ________________________ actions.

The **future tense** is used to reference __________________ or _________________ that have not yet _____________ ________________.

Adjectives and Adverbs

Adjectives are used to describe ___________________ or _____________________.

Adverbs describe __________________, _________________________, and other _________________.

Prepositions

Prepositions show __________________________ among __________________, __________________, _____________________, _______________________, and ______________________.

Prepositional phrases are word groups made up of a ___________________________ and the ____________________ **of the preposition**.

Conjunctions

A **conjunction** is a part of speech that __________________________ two or more ____________________, _____________________, or __________________________ in a sentence.

Coordinating Conjunction:

Subordinating Conjunction:

Correlative Conjunction:

Interjections

Interjections are ______________________ or short ___________________ of words that show _________________________ or _________________________.

Subject and Predicate

A complete sentence must _______________ include a ________________ and a ___________________.

Subject Nouns and Pronouns

Subjects are usually stated _________________ as nouns or pronouns.

Write your own sentence with a subject noun:

Write your own sentence with a subject pronoun:

Subject Placement

The subject does not always have to be the ________________ word in the sentence.

Write your own example of a subject that is not the first word in the sentence:

Implied Subjects

Sentences giving the audience a ___________________ or ___________________ contain an **implied subject**.

Write your own sentence with an implied subject:

Compound Subjects

Subjects can also be compound. In these instances, the subjects are linked by ________________.

Write your own sentence with a **compound subject**:

Subjects vs Objects

Although nouns and pronouns often function as subjects, they also often function as objects. When looking for the subject of a sentence, you can rule out the following types of ____________________:

1.
2.
3.

The function of a pronoun determines its **case**. There are three pronoun cases:

1.
2.
3.

Predicates

Like nouns and pronouns, verbs can take on a variety of forms that serve different purposes in sentences.

Word Groups

Remember that predicates are often made up of a ________________ _____________ + a __________________ ________________.

Write your own sentence with a predicate made up of a helping verb and a main verb:

Compound Verbs

Write your own sentence with a compound predicate:

Verbs with Other Functions

Verbs can also be used as ________________ or _______________ in a sentence.

On Your Own

Identify the subject and the main verb in each of the following sentences.

The plane finally landed after about two hours of circling.

Before starting the baking process, you should measure all your ingredients.

Jane Austen's *Pride and Prejudice* was made into a popular movie in 2005.

Steps for Locating the Subject and Predicate

1.

2.

3.

Complete Thought

When a group of words expresses a **complete thought**, the group of words is ____________________ ____________________. If a group of words contains a subject and a verb but does not express a complete thought, it is known as a ______________________ ______________________.

On Your Own

Read the following paragraph and identify the incomplete sentence.

> Ever since Sharrell visited the Museum of Modern Art in New York City. She has been re-inspired to create unique pieces of pottery. She took her first pottery class in high school and quickly picked up the craft with an innate sense of creativity. Her ability to shape the clay was natural, but her skills were further sharpened by taking advanced pottery classes in the following years. After high school, lack of free time put a stop to her art. However, her passion for art has been rekindled and she now seeks to make pottery a priority.

Lesson Wrap-up

Test Yourself

In the following sentences, identify the subject(s) (some may be implied) and add the correct punctuation. Identify whether each sentence is declarative, interrogative, imperative, or exclamatory. Check your work using the Answer Key in the back of the book.

Did you know that Mount Everest divides China and Nepal

Our semester begins in late August and ends in early June

Please submit your employment application on our website portal

I can't believe you are moving to Germany

Key Terms

Beside each definition, write the corresponding term.

______________________: A sentence that communicates emotion or surprise

____________________: A word that describe a noun or a pronoun

____________________: A verb tense used to describe events or actions that have not yet taken place or to describe plans or instructions

____________________: The simplest form of a verb, free of alterations for tense, number, and point of view

____________________: A word that completes the meaning of a prepositional phrase

____________________: A word that takes the place of a noun in a sentence

____________________: A word group that contains a subject and a verb

____________________: A part of speech that describes a verb, adjective, or another adverb

____________________: A part of speech that connects two or more words, phrases, or clauses in a sentence

____________________: A punctuation mark that conveys emphasis or strong feeling

____________________: Conjunctions that connect two or more similar ideas and always appear in pairs

____________________: A sentence that makes a general statement that informs the reader

____________________: A word that receives the action of a verb

____________________: A sentence that asks the audience a question and usually ends with a question mark

____________________: A pronoun that functions as an object in a sentence

____________________: A conjunction that connects words or word groups of equal importance in a sentence

____________________: A pronoun used to refer to the speaker of the sentence

____________________: When a group of words is logically finished

____________________: A word that changes the form of the main verb so that it grammatically fits in the sentence

____________________: A sentence that makes commands and often addresses the implied subject

____________________: The subject of a sentence when the speaker is talking directly to the audience

____________________: A pronoun that does not rename a specific noun

____________________: A word group that contains a subject and a verb and expresses a complete thought

______________________: A sentence that contains at least one subject and one predicate and expresses a complete thought

______________________: A sentence that tells the audience about a question and usually ends with a period

______________________: A word or group of words that adds emphasis or emotion

______________________: A sentence that asks a question

______________________: A verb that links the subject to a description

______________________: The verb that expresses the primary action or state of being of a subject in a complete sentence

______________________: A sentence made up of one independent clause

______________________: The categories that classify each type of word found in a sentence

______________________: A verb tense used to report on an event or reflect on a past experience

______________________: A punctuation mark used to end sentences that make a statement or relay information

______________________: A pronoun that renames a specific person, animal, object, or place

______________________: A pronoun that shows possession of another noun

______________________: The part of the sentence that states what the subject is doing

______________________: A word that receives the direct object

______________________: A group of related words that starts with a preposition and ends with a noun or pronoun

______________________: A word that shows a relationship among people, places, things, events, and ideas

______________________: A punctuation mark used at the end of a sentence that asks a question

______________________: The form a pronoun takes based on its function—subjective, objective, or possessive—in a sentence

______________________: The person, place, thing, event, or idea a sentence is about

______________________: The form of a pronoun that functions as the subject in a sentence

______________________: A word that represents an action, relationship, or state of being

Answer Key

CHAPTER 0R

0R.2 Determining Your Personal Learning Styles

TEST YOURSELF

1. T
2. F
3. T
4. T
5. F
6. F
7. T
8. T
9. F
10. Active
11. Intuitive
12. Sequential

0R.3 Stress Management

TEST YOURSELF

1. F
2. T
3. F
4. F
5. T
6. T
7. T
8. External
9. Internal
10. Semester exam

0R.4 Keeping Yourself Organized

TEST YOURSELF

1. F
2. F
3. F
4. T
5. T
6. T
7. T
8. T
9. F
10. T

0R.5 Time Management

TEST YOURSELF

1. T
2. F
3. T
4. F
5. F
6. 10 hours with friends
7. At least 6 hours studying
8. 24 hours of studying
9. Top-down strategy
10. Multitasking

0R.6 Study Strategies

TEST YOURSELF

1. F
2. T
3. F
4. T
5. F
6. F
7. T
8. T
9. F
10. T

0R.7 Test-Taking Strategies

TEST YOURSELF

1. F
2. T
3. T
4. T
5. F
6. T
7. F
8. T
9. T
10. F

0R.8 Taking Advantage of Campus Resources

TEST YOURSELF

1. Student Services
2. The Library
3. Disabilities Services
4. Student Services
5. The Writing Center
6. The Academic Success Center
7. The Library
8. The Writing Center
9. The Library
10. Disabilities Services

CHAPTER 1R

1R.1 Understanding Purpose

TEST YOURSELF

1. I
2. F
3. F
4. I
5. F

CHAPTER 2R

2R.3 Identifying Organizational Patterns

TEST YOURSELF

1. c
2. e
3. a
4. b
5. d

CHAPTER 3R

3R.1 Writing a Paragraph

TEST YOURSELF

1. F
2. T
3. F
4. F
5. T
6. F
7. F
8. F
9. T
10. T

3R.2 Writing a First Draft

TEST YOURSELF

1. F
2. F
3. T
4. T
5. T
6. F
7. T
8. F
9. T
10. T

3R.4 Identifying the Characteristics of Sentences

TEST YOURSELF

- Whenever the four of us go shopping, I always spend too much money. (period)
- Did you know the man down the street who passed away? (question mark)
- Marcus and Steve joined our group and helped us raise money. (period)
- Every time you get behind the wheel of a car, put your seatbelt on. (period)
- I'm *shocked* by how much you look like Steve Carrell! (exclamation point)

CHAPTER 5R

5R.3 Active Reading Strategies

TEST YOURSELF

1. N
2. Y
3. Y

5R.5 Finding Meaning Through Context

TEST YOURSELF

1. sad (answers will vary)
2. antonym
3. scared (answers will vary)
4. example
5. unlike

5R.9 Identifying Purpose and Tone

TEST YOURSELF

1. E
2. I
3. E
4. P
5. R
6. I
7. T
8. F
9. T
10. T

CHAPTER 6R

6R.5 Emphasizing Words or Phrases

TEST YOURSELF

- Even though you have been busy, you still should've called me.
- Twirling gracefully in the fall breeze, the leaves reminded me of little ballerinas.
- After careful consideration, I have decided not to take the job in London.
- Whenever my mother-in-law comes around, I feel like I have been judged on my child-rearing and my housekeeping.
- In order to be considered for this promotion, employees must submit their Request of Interest form to Human Resources.

CHAPTER 8R

8R.1 Considering Style

TEST YOURSELF

1. F
2. F
3. F
4. I
5. F
6. I
7. I
8. F

8R.2 Considering Visuals

TEST YOURSELF

1. F
2. F
3. T
4. F
5. T

CHAPTER 9R

9R.1 Nouns

GRAMMAR PRACTICE 1

- One lucky student will win this awesome bike. (student = subject)
- My sister always gets good grades. (sister = subject)

GRAMMAR PRACTICE 2

- The chicken ate a bug. ((bug = direct object)
- The farmer planted his crops. (crops = direct object)

GRAMMAR PRACTICE 3

- My nephew hid under the table and wanted me to find him. (table = object of the preposition)
- We went down the river and swam in the deep water. (river = object of the preposition) (water = object of the preposition)

GRAMMAR PRACTICE 4

- The mail truck has arrived! (mail = adjective)
- I can't believe the cat knocked over his own food bowl. (food = adjective)

9R.2 Pronouns

GRAMMAR PRACTICE 1

- After soccer practice, I made sure to drink plenty of water.
- You aren't really a true Jacksonville Jaguars fan unless you get upset when they lose!

GRAMMAR PRACTICE 2

Singular Pronouns

- I was in the middle of cooking steak and potatoes when my friend invited me out to dinner.

Personal Pronouns

- The group decided that we would go for a walk after dinner; our stomachs needed help digesting.

GRAMMAR PRACTICE 3

First-Person	Second-Person	Third-Person
What is up with me these days? I have been so scatterbrained lately that my friends are concerned.	What are your plans later on today? If you would like, we can go shopping.	She has been doing such a great job at work that they have promoted her to manager.

GRAMMAR PRACTICE 4

1. his
2. her
3. their

GRAMMAR PRACTICE 5

Answers will vary.

GRAMMAR PRACTICE 6

- At the time of the election, many people were unhappy with both candidates, which led to political unrest.

TEST YOURSELF

1. I (subjective), my (possessive)
2. They (subjective), you (objective), me (objective)
3. You (subjective), their (possessive)
4. My (possessive), that (relative)
5. Whomever (indefinite)

9R.3 Basic Verb Types and Tenses

GRAMMAR PRACTICE 1

- will
- am

9R.6 Adjectives and Adverbs

GRAMMAR PRACTICE 1

- The elderly man limped painfully after his surgery.
- I shop at that store often.
- Mom hid her jewelry carefully.

9R.7 Prepositions

GRAMMAR PRACTICE 1

Answers will vary.

9R.8 Clauses and Conjunctions

TEST YOURSELF

Coordinating Conjunctions

1. I wanted ham, but we didn't have any, so we went out to eat instead.
2. We had ice cream and cake at the birthday party, but not everybody wanted some.
3. He and I will be getting married next year, so we have started planning now.

Subordinating Conjunctions

1. Ever since I started exercising, I have felt a lot better.
2. After snacking all day, I couldn't eat dinner.
3. In spite of our differences, we have remained friends since first grade.

Coordinating Conjunctions Joining Prepositional Phrases

1. We went over the mountain and through the woods.
2. I went through the yard and into the house.

9R.12 Spelling Commonly Confused Words

TEST YOURSELF

1. accept
2. affect
3. its
4. past
5. loose
6. choose
7. than
8. there
9. They're
10. too
11. to
12. weather
13. whose
14. You're
15. between
16. come
17. fewer
18. lie

CHAPTER 10R

10R.1 Identifying the Characteristics of Sentences

TEST YOURSELF

- Did you know that Mount Everest divides China and Nepal? (question mark) (Mount Everest = subject) (interrogative)
- Our semester begins in late August and ends in early June. (period) (semester = subject) (declarative)
- Please submit your employment application on our website portal. (period) (you = implied subject) (imperative)
- I can't believe you are moving to Germany! (exclamation point) (I = subject) (exclamatory)

Notes

Notes

Notes

Notes

Notes

Notes

Notes

Notes

Notes